THE LOOKOUT

❖

*National Maritime Museum;
'Old Ironsides'
Whaling Museum;
Steam Heritage;
Nelsonia?*

National Maritime Museum

At the National Maritime Museum at Greenwich, London, the very popular exhibition *The Wreck of the Titanic* will remain open until 1 October 1995. In remembrance of the ship and those who were lost, the museum has established a Titanic memorial garden, incorporating a stone memorial, in its grounds.

A major new exhibition, *Nelson – Life and Legend*, will open at the Museum on 21 October 1995. It marks the beginning of 'The Nelson Decade' (to which reference made in *MS 91*). The exhibitio comprise an examination of th and the legend, supported by a bination of interactive and a visual displays, more than hundred artifacts, including th form in which he died and the ket ball which killed him, and a display of his personal letters.

'Old Ironsides'

'Old Ironsides' and the US Navy: 200 Years in Scale Models (1797-1997) is the title of a competition and exhibition to celebrate the 200th anniversary of the launch of the frigate USS *Constitution* being organised by the USS Constitution Museum in Boston, MA.

The competition is open to all completed models of American naval vessels that were or are in service during *Constitution*'s 200-year career built by individual, professional or amateur modellers. Models of foreign vessels, and of other vessels, may be eligible if they have a direct relationship to *Constitution*, or re-
vari-
:lude
t and
cut-
bot-
rate

pac-
ket, containing entry forms, rules, list of special awards, brief bibliography, and a list of *Constitution* plans available from various sources, can be obtained from: 'Old Ironsides' and the US Navy Scale Model Show, USS Constitution Museum, PO Box 1812, Boston, MA 02129, USA. Entry forms have to be in by 1 May 1997. There is a registration fee of US$10.00 per model. The exhibition will run from 1 July 1997 to 2 November 1997.

The half-scale model of the whaling ship *Ladoga*.
Photograph: courtesy of New Bedford Whaling Museum

Whaling Museum

In his article in *MS 92* on building a model of a beetle whaleboat, Pat Earle referred to the new Bedford Whaling Museum. Among the many interesting exhibits which he saw in this museum was a full size beetle boat, which showed the tough and strong construction adapted to withstand the rigours of their use. Built from locally grown wood, they lasted on average just two seasons when carried on whalers. Another exhibit was an 80ft half scale model of the whaling barque *Ladoga*, which was fitted out below decks to quarter scale. There were a number of models of whaling ships. Two skilled scrimshaw engravers were demonstrating the art, working on imitation whalebone. The skeleton of a hump back whale was impressive for its awesome size. For those interested in research the museum has over 1100 log books from ships engaged in whaling.

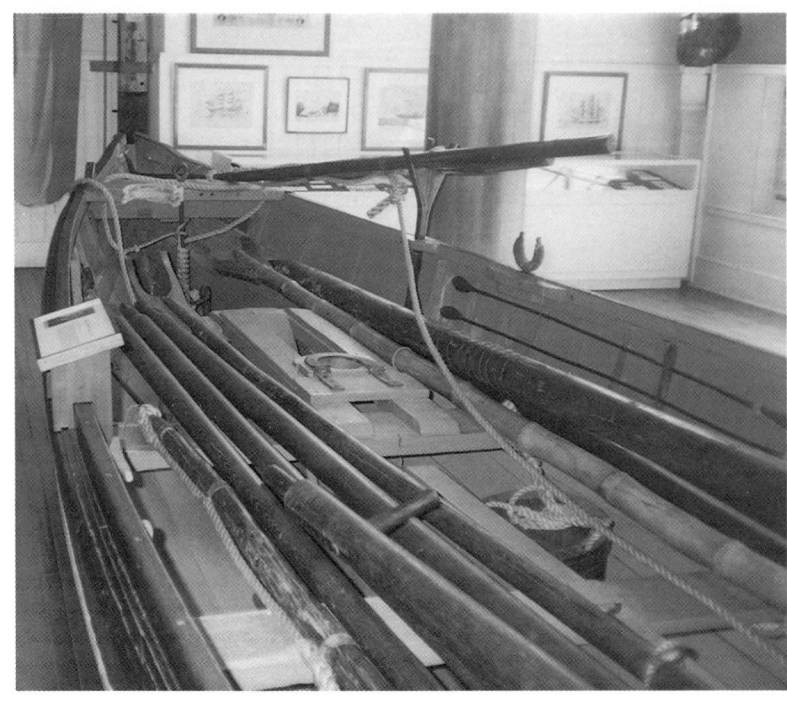

Steam Heritage

The *Steam Heritage Museums & Rally Guide 1995/6* contains a wealth of information about preserved railways, industrial works, transport, aircraft, ships and military equipment in this country. As well as listing all the preserved railways, there are sections containing details of relevant national associations; museums, steam centres and sites of interest; canal, ship, aircraft and military vehicle museums; publications and periodicals of interest; steam operated miniature railways and a 1995 rally and events diary. This guide is a real eye-opener to the vast steam heritage in this country. The ship section contains some interesting entries. The Guide can be obtained from newsagents and preservation centres, price £2.50, or direct from the publishers, *Tee Publishing, The Fosse, Fosse Way, Radford Semele, Leamington Spa, Warwickshire CV31 1XN*, for £2.95 including postage.

Beetle whaleboat. Note harpoons ready in crutch, killing lances below starboard gunwale, pulling and steering oars on thwarts.
Photograph: Pat Earle

Model of a whaling ship in the museum.
Photograph: Pat Earle

Nelsonia?

Recently, *The Times* carried a report in which it was stated that Nelson's arm was preserved in a museum in Santa Cruz, Teneriffe. A follow-up letter from the Nelson Society pointed out that his arm was amputated at sea after the Teneriffe incident, and doubtless disposed of overboard. Well, well!

The Dundee Steam Whalers

by I M Fleming

In *MS* 31 (March 1980) I wrote about the Antarctic research ship *Discovery* (1901). Working backwards, I wanted to find information on the type of ship from which *Discovery*'s form was derived, namely, the auxiliary screw whalers built at the Panmure yard, Dundee, Scotland, by Alexander Stephen & Son between 1859 and 1884.

We must remember that these activities belonged to a society that depended heavily on animal products. The oil from seal and whale blubber was used for lighting, lubrication, and some industrial processes, while 'whalebone' (not actually bone, but the baleen plates from the whale's mouth), which fetched a high price, counted among its applications the notorious 'waspwaist' ladies' corset. The meat was not used. Eventually, however, the Greenland whaling trade failed by its own success, as its quarry, the right whale *Balaena mysticetus*, was reduced nearly to extinction soon after the turn of the century.

Dundee owners had been sending ships to the whaling grounds since 1754. Their first steam whaler was the *Tay*, built in 1850 by Martin of Dundee for the India trade, engined and strengthened for whaling in 1857. Her success prompted the building of specially-designed steam whalers beginning with the *Dundee* and Stephen's *Narwhal*, both completed in 1859; the origin of *Dundee* and *Wildfire* (1860) is unknown. Of the surviving sailing whalers, only *Jumna* was given an engine, in 1863. Steam increased costs, but brought worthwhile bene-

The steam whaler *Balaena*.
Photograph: Valentine Collection, St Andrews University Library

America (left) lying alongside the new *Discovery* while being fitted out at Dundee.
Photograph: M'Intosh Collection, St Andrews University Library

fits, as steamers were able to push through pack ice to the best whaling grounds and work there more safely and for a longer season than the old sailing ships.

Following *Narwhal*, Stephen's yard turned out about one steamer a year (twenty-six in total) until 1884, ending with the renowned *Terra Nova*. Many were built for owners outside Dundee, notably those in the Newfoundland sealing trade. Although the same ships could be used for sealing and whaling, the two eventually diverged as separate types, distinguished largely by engine and mast layout, as will be discussed below.

Another Scottish port, Peterhead, was Dundee's main rival, using sailing ships some of which (*Active, Windward*) were later fitted with engines, although less powerful than their Dundee cousins. In 1867 Hall of Aberdeen built the *Eclipse*, on similar lines to Stephen whalers, for Peterhead owners, followed in 1873 by the handsome, gunport-painted *Hope*. Of these, *Active, Windward* and *Eclipse* later came into Dundee ownership, along with secondhand Norwegian and German ships including *Balaena, Diana, Vega* and *Morning*, to see out the last years of the Arctic whaling after the original Dundee companies had given up as the whale grew scarce and sold their surviving ships, some of which continued sealing for many years. In 1892-93 an expedition comprising four ships from Dundee sailed to the Antarctic to hunt Southern whales, but soon found out that these swam too fast to be caught by men in rowing boats. Antarctic whaling required a totally different kind of ship, which the Dundonians did not trouble to develop.

Technical description

This description is of a Stephen-built whaler or sealer; some other Dundee-owned ships are described in the Lloyd's surveys mentioned below.

Iron being unable to withstand the pressure of ice, these ships were built of wood, with enormous strength. Keel, stempiece, sternpost, frames and deck beams were of Baltic oak, of great thickness. In the *Bear*, the frames were 24in x 15in with 30in spacing; deck beams between 8in and 12in square with 5ft spacing. On top of the 6in oak planking, secured with iron bolts, was laid a second skin: yellow pine on the bottom (absent on the earliest ships where the bottom was left unsheathed), tough iron-bark 2½in thick on the sides for a few feet above and below the waterline. On at least the *Terra Nova* this sheathing continued up to the covering board, but on most known examples it ended 2ft or 3ft lower, resulting in a definite ledge visible on the ship's side. (*Eclipse* and *Hope* had yet a third layer from the bow to the foremast.) Adding the ceiling of 5.5in, the sides were thus up to 29in thick. The raked bow was further sheathed with ½in iron plates and supported internally with thick diagonal shores. All this was necessary to withstand the 'nipping' force of ice floes, which was strongest around the waterline, and to enable the ship to force through pack ice by riding up on to it and using the dead weight of the bow to crack it open.

The hull was short, to facilitate

swift turning when avoiding floes. Length, typically about 160ft, ranged from the *Wolf* I at 131.5ft to the *Arctic* II at 200.6ft. Breadth was about 29ft and depth was about 18ft to 19ft irrespective of length. Within these registered dimensions the hull was fairly capacious, with rather sharp bows and a smooth run aft to give a good flow to the screw and rudder, but a full midsection, with a little tumblehome, and a broad curved transom.

Internally, watertight transverse bulkheads were provided near the bow and stern and between the hold and machinery spaces. The hold, roughly between the fore- and mainmasts and reaching up to the main deck, contained iron tanks for whale and seal blubber. Between main and mizzen were the boiler and engine, simple and rugged machinery of about 70hp nominal or 300hp indicated. In some of the barquentine-rigged ships the engine was further forward, with the funnel before the mainmast so that a boom mainsail could be carried; such were the *Nimrod* and *Bear*.

Sails were retained, in case of engine breakdown or damage to the propeller, to save coal, and to avoid the noise of the propeller giving warning to the whales. In the first few ships, as evidenced by the *Narwhal* model mentioned below, square sails were carried on all masts, the single topsails on the fore- and mainmasts having (though perhaps not originally) roller-reefing gear; this model also has gaffs on the fore and mizzen masts. Later ships dispensed with square sails on the mizzen and in many cases the main as well. Basically, with exceptions (*Aurora*, though a barquentine, had the funnel abaft the main;

Another view of *America* alongside *Discovery* while being fitted out.
Photograph: M'Intosh Collection, St Andrews University Library

Arctic II was barque-rigged but had her funnel amidships; *Terra Nova* was a sealer- rigged like a whaler) these two types emerged: barque-rigged whalers with engine aft, to give room on deck between the fore and main for processing the catch, and sealers with barquentine rig and engine amidships.

The screw, two-bladed and about 10ft diameter, could be lifted into the hull, while the rudder could be unshipped over the transom.

The *Narwhal* model shows one continuous weather deck with only a raised grating at the heel of the bowsprit; by the time of *Esquimaux* a proper forecastle deck had been added, while the later ships, especially sealers, also had a raised quarterdeck.

Whales were pursued by small boats, not directly by the parent ship. Eight were carried, each with a harpoon gun in the bow and a crew of fourteen. To this day, a double-ended open rowing-boat is referred to as a 'whaler' – now you know why! – although perversely the boats used on Peterhead whalers had transom sterns. Sometimes extra boats were carried when sealing, as noted later.

Sources

The intending modeller of any Dundee whaler will soon discover that there is remarkably little reference material, considering their number. There are *no* surviving builders' plans, either of the Stephen ships (records of which were lost in a fire) or of any of the secondhand whalers of the last period, except perhaps *Eclipse*. (*Eclipse* and *Hope* may be represented in the Hall collection at the National Maritime Museum). However, there are dockyard plans in the NMM of the two Stephen sealers taken into Government service: *Terra Nova*, acquired in 1903 for the relief of Scott's *Discovery*, is shown by plans 8049, general arrangement; 8050, rigging profile; 8051, decks; 8052, midship section. (There are also plans of her in Broughty Castle Museum.) *Bloodhound II*, built in 1873, was bought in 1875, modified for Sir George Nares' expedition towards the North Pole, and renamed HMS *Discovery*. She is shown by NMM plans 7396, lines; 7397, hull profile; 7400, plan with inboard elevation of bulwark; 7401, stays and sails; 7594, midship section. Note that this ship is referred to below as *Bloodhound*, to avoid confusion with the 1901 *Discovery*.

In Harold Underhill's range of plans now marketed by Brown, Son & Ferguson Ltd, Glasgow, the set entitled 'Barque-rigged auxiliary whaler' is based on Scott's *Discovery* (regrettably it contains some inaccuracies), and so is not really useful for whalers at all.

There are a few relevant photographs in the collections of Broughty Castle Museum (in Broughty Ferry, near Dundee), Dundee Central Library, St Andrews University Library, the NMM, and the Shetland Islands Museum. (There must be more, at least of the sealers, in Canadian archives; perhaps a reader on that side could enquire?). Lubbock's book contributes a handful more. Some of the photographs may not have been correctly identified or dated, and ships of such longevity (*Neptune, Terra Nova* and the erstwhile *Eclipse* all survived until 1943) are liable to change their appearance from time to time; but the camera does not lie, and is the source on which greatest reliance has to be placed.

An article in the *Illustrated London News* for 16 March 1861 includes an engraving (p242) entitled 'Dundee screw whalers', showing the *Tay, Dundee, Wildfire, Narwhal, Polynia* and *Camperdown* leaving the Tay; it is the only pictorial evidence for any of them. Impressive though it is, an artist who shows rows of portholes on non-passenger ships and elevates catheads to the same steep angle as davits has to be suspect. If, even so, you want a print, one can be obtained from the ILN Picture Library, Sea Container House, 20 Upper Ground, London SE1 9PF.

Lloyd's Registers give the hull dimensions and tonnage for each ship (as does Lubbock for all the Stephen ships), and in most cases – including all the sailing whalers – this is the only information of any kind. The dimensions given are length, measured from the fore side of the stem to the aft side of the sternpost along the uppermost continuous deck, and following the curve of its sheerline; breadth to outside of frames, and depth from deck beams to floor. Thus the thickness of planking and sheathing should be added to give the extreme breadth, and the depth of keel and thickness of floor timbers taken into account, for modelling purposes.

The NMM has Lloyd's surveys of the *Active, Chieftain, Maud, Tribune, Vega* and *Windward*. These interesting documents give information on the hull construction and scantlings, but their use to the modeller is limited. We would like to have the spar dimensions, for example, but learn only that they were 'sufficient in size and length'. It is a pity, too, that no Stephen ship is represented.

A fine and authoritative model of *Eclipse* is displayed in the Royal Museum of Scotland, Edinburgh, and was described by its maker James Gray in *The Mariner's Mirror* No 23, 1937, p446-48. Other models of *Eclipse* are in the maritime museums at Aberdeen and Peterhead, and there are models of the *Hope* in Broughty Castle and Peterhead museums. Broughty Castle Museum also has models of three Dundee whalers and of the engine fitted in the *Active*. One, the *Tay*, is probably misidentified. The *Narwhal* is represented by a fascinating contemporary model thought to have been made by her carpenter in 1870; though crude, it is a useful source for deck layout and fittings as well as rig. The model has excessively long spars and a minute funnel, which seems like a deliberate joke, perhaps reflecting its maker's relative opinions of steam and sail. The third model, of *Terra Nova*, is modern, and quite exquisite.

Books and periodical articles afford some help, sometimes in a roundabout way. In contrast to the well-documented American whalers of the same period, almost nothing apart from Lubbock has been written on the Scottish ships in their capacity as whalers.

Carvel's history of Alexander Stephen & Sons is also useful, and corrects Lubbock in some details,

America in Davis Strait, c1901.
Photograph: Dundee Art Galleries & Museum, Dundee

particularly in identifying which ships were built as sealers. The booklet by Henderson (formerly Curator of Broughty Castle Museum, who more than once gave up his lunch-hour to answer my questions) contains a table of all known Dundee whaling ships and descriptions of the museum's collections.

However, these ships were not only used for whaling. Built to pursue their trade at the ends of the earth, they were much in demand for Arctic and Antarctic exploration. Such expeditions – being of more interest to contemporary publishers than whaling – have left accounts which include some descriptions and illustrations of Stephen-built ships. Nares' *Discovery* has already been mentioned. *Neptune, Proteus, Bear* and *Thetis* all have a place in the tragic story of Greely's expedition to Lady Franklin Bay in 1881-84. The last two subsequently served with the US Coast Guard, and *Bear* sailed with Byrd's expedition to the Antarctic in 1933-35. *Terra Nova*, after assisting Scott's first Antarctic expedition, gained greater glory in his fatal South Pole journey (1910-12). Shackleton chose the little *Nimrod* for his first Antarctic expedition (1908) and *Aurora* (along with the ex-Norwegian sealer *Endurance*) for his second (1914-16). *Aurora* had already been to the Antarctic with the Australian Douglas Mawson (1911-14).

The literature of these voyages makes at least an absorbing digression, and yields some very useful information, such as the profile drawing, photographs and technical description of the *Bear* in Wead. But most of the ships concerned were sealers, while I wanted principally to research the whalers. Fortunately, I had found enough information about one of them to attempt a reconstruction.

Esquimaux

The *Esquimaux* seems a natural choice, there being several photographs in existence, and one painting. She is a handsome ship, fairly simple to model, and full of interest.

Alone among the Stephen-built vessels, *Esquimaux* was not completed as a whaler, and it was not until her second year (1866) that her engines and outer planking were fitted to equip her for the trade. But she was always intended as such, being owned from the start by the Dundee Seal and Whale Fishing Company, until its liquidation in 1895. She is not a converted merchantman, but a true and basically typical whaler.

Esquimaux's particulars are: registered dimensions 157.3ft x 29.5ft x 19.3ft; 593 tons gross, 466 net; 70hp engine auxiliary to sail; built 1865 by A Stephen, Dundee. In 1876 she was given new boilers. She was sold in 1895 to a syndicate who formed a new Dundee Arctic Fishing Company; in 1898-99 she was used by A Barclay-Walker as an 'Arctic yacht' for a hunting and exploration voyage. After the 1900 season she was acquired by US owners. Renamed *America*, she sailed with the two Ziegler expeditions: to Franz Josef Land in 1901 under the command of Evelyn B Baldwin, and a second in 1903-5 commanded by Anthony Fiala, who wrote an account of both voyages. During the second the old ship was abandoned and the crew rescued by the *Terra Nova*.

The painting of *Esquimaux* is reproduced in Lubbock, facing p394, and probably shows her at a fairly early date.

The photographs available are:

A. Reference 1969-117-1 in Dundee Museums (correspondence should be addressed to Dundee Art Galleries and Museums, Albert Square, Dundee DD1 1DA). A port broadside view, thought to have been taken off Newfoundland in 1894, of the ship under sail but fast in ice, with most of her crew standing on the ice. Although faint, as the dark areas have faded on what must have been a glass plate, the picture is particularly useful for the arrangement of boats and for the sails. The identification is certain, as the ship's name appears near the bow.

B. Wilson 912 in Dundee Central Library (Wellgate, Dundee DD1 1DB) shows the paddle-steamer *Teazer* in the foreground, with *Esquimaux* moored against the har-

Contemporary model of the *Narwhal* in Broughty Castle Museum, Broughty Ferry, Scotland.
Photograph: author

bour wall behind, shown in a starboard view. This picture is small and cluttered, but a good magnifying glass confirms the name and various details. I used it in comparison with later views to check what alterations were made when the ship became *America*: for example, the Eskimo figurehead, which was applied later and was not original. The drawings given here are intended to show her at this earlier period.

In 1901, as the new *Discovery* was being fitted out at Dundee, the renamed *America* lay alongside her. Several photographs taken at that time have been found, the next two as tiny sepia prints pasted in scrapbooks in the St Andrews University Library archives:

C. From Album 5, M'Intosh Collection (University Library, St An-

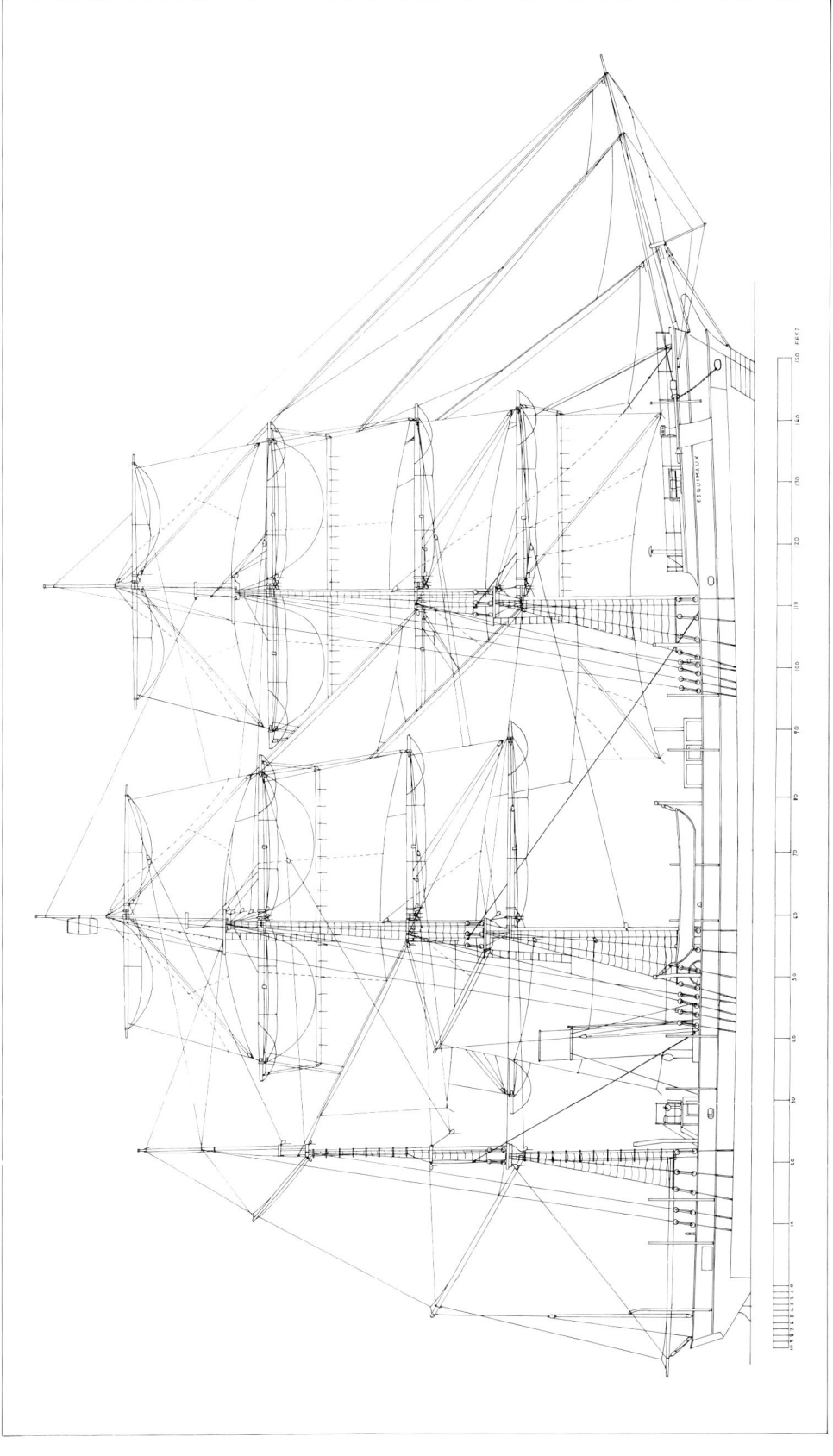

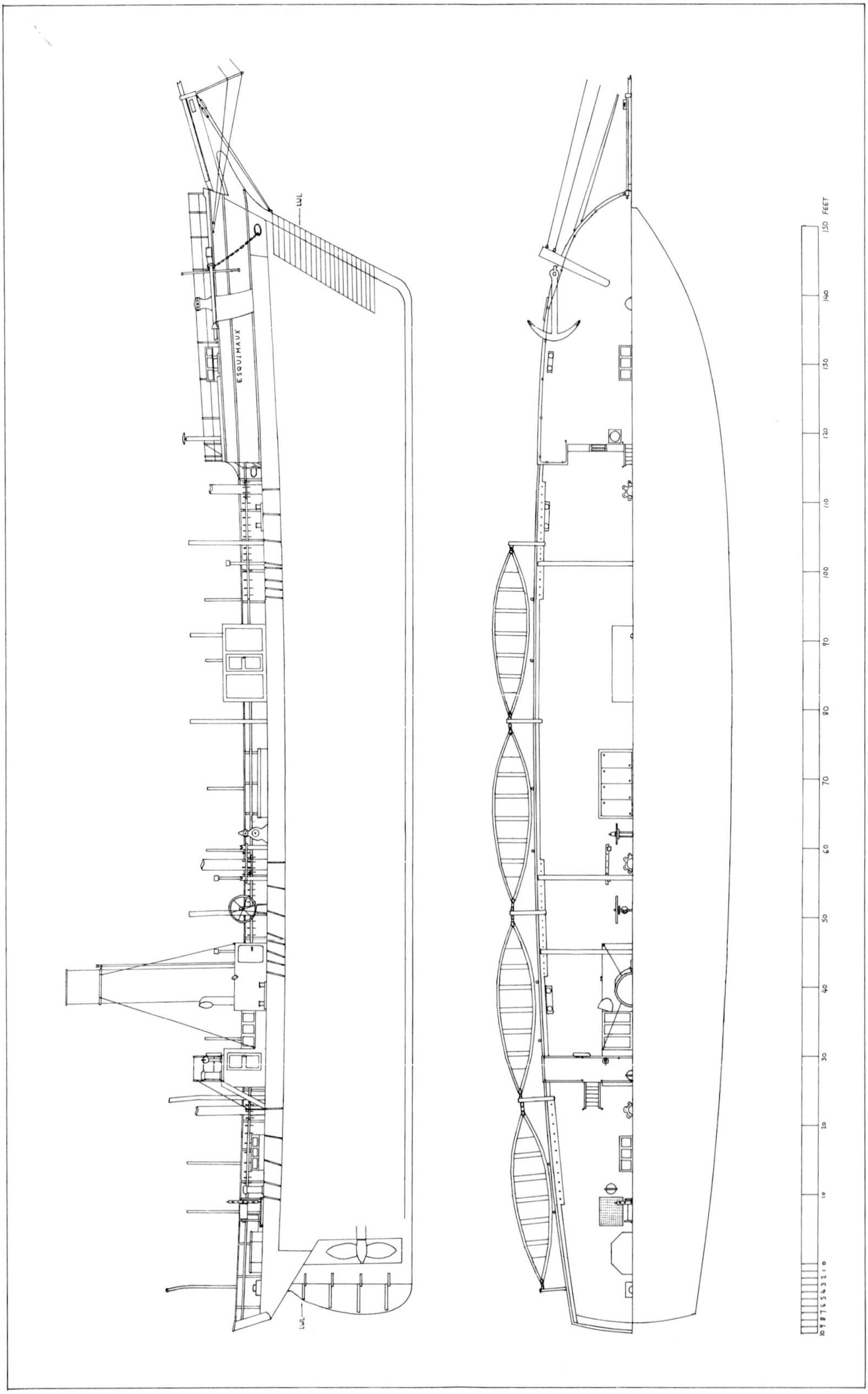
LWL
ESQUIMAUX
LWL
150 FEET

The *Terra Nova* (1884).
Photograph: author's collection

drews, Fife KY16 9TR) comes a faint and badly marked bow view of the two ships, useful for fo'c'sle and mast detail.

D. The same collection's Album 3 has the picture that made this reconstruction possible. *America* is shown in front of *Discovery* in such a way as to enable the sheerline, mast and funnel spacing and rake, and many details to be compared with known measurements on the latter ship. It also shows clearly the large deckhouse that was added on the after deck to accommodate the hunters on later voyages – again, this is not original and is therefore not shown on the drawings.

E. Wilson 766 (Dundee Central Library), dated 19/4/1901, again shows *America* alongside *Discovery*, with another ship between, in a starboard bow view.

F. Wilson 765, dated 25/5/1901, shows *America* alone, now moored to the wall of the same basin, in a starboard side view. It is more useful than the previous photo for showing the rigging, which stands out better against the sky, although in both photos the topgallant yards have not yet been fitted and the rigging is incomplete.

G. Dundee Museums reference 1975-351-109, misidentified as *Eclipse* in their publications, is a nice view of *Esquimaux/America* at sea, under steam but not sail, with the crow's nest hoisted and boats swung out ready for a chase. The identification is confirmed by the presence of the aft deckhouse shown on photo D.

[Note: it has not been possible to reproduce all of these photographs. Editor].

Establishing the dimensions

I began a preliminary drawing by marking the registered length, less a small allowance for the curve of the sheer, and constructed the sheerline with reference to the NMM drawing of the *Bloodhound*, Wead's profile drawing of the *Bear*, and comparison with the later *Discovery* in photograph D. As registered depth is an internal measurement it is difficult to use directly, and so I placed the keel 1ft below *Bloodhound*'s as this is the difference between their depths. I made the rake of the stem equal to that on *Bloodhound* and *Bear*, which looked right compared with the photographs, and thus fixed the keel length. The bow is similar in form to *Bear*'s, with the sheathing of iron plates. At the other end, I reduced the angle of the transom as suggested by photographs D and G, but otherwise followed *Bloodhound*'s form.

With the hull outline determined, the next stage was to place the masts and funnel. None of the photographs is an exact broadside (and there would be distortion towards the ends even then), so measurements

The *Nimrod* (1866).
Photograph: Trustees of the National Maritime Museum, Greenwich

taken from them have to be corrected for perspective. This is where photographs D and E, taken from abaft and before the beam respectively, are a real help. I measured on photograph D the *apparent* ratio of fore-main and main-mizzen distances on the *Discovery*, compared it with the known actual figure, and thus obtained a factor for correcting the same ratio as measured on *America* in the same picture, the two ships being moored parallel. In the same way, ratios for the distances between mizzen and sternpost, and between knightheads and mast were calculated, and all were cross-checked by using the same method applied (less satisfactorily, because of the closer view and greater angle) to photograph E. The results tallied nicely with the painting in Lubbock, which is a broadside view.

Establishing the height of the masts is more difficult, as with the low camera angle it is impossible to compare directly the mast heights on *Discovery* and *America*, although the rake of the masts and funnel can be compared. From photographs D and G, *America*'s fore- and mainmasts seem to be vertical or only very slightly raked, the mizzen a little more (and the funnel more again) – making her look rather old-fashioned next to the sleeker *Discovery*. So, to fix the mast heights, recourse has to be made to counting ratlines, which usefully were normally set 15in apart. Counting on the clearest photographs, the foreyard is level with the 22nd ratline above the sheerpole, and its position below the foretop (by comparing with *Discovery* on photograph C) is about 4ft. Adding some 4ft from the bulwark to the top of the deadeyes gives a length along the shrouds of 4+(22 x 1.25)+4 = 35.5ft. This is the hypotenuse of a right-angled triangle whose base is half the beam at that point, about 12ft. Drawing the triangle and measuring the resulting vertical gives 33.5ft for the height of the mast from the level of the toprail to the foretop. By similar means the length of the doubling and of the topmast can be ascertained, but the topgallant, having no ratlines, poses a problem. On the main topgallant there is a wire ladder from the crosstrees to the hounds, by which the crow's nest was reached; the spacing of its sixteen horizontal battens gives its length as 20ft. Correcting for its slope, the mast is 19ft to the hounds, leaving only the necessity of estimating by eye the height from there to the truck as about another 12ft. The total height of the ship from keel to main truck of about 136ft is some 17ft more than on *Discovery* (whose masts were thought to be too short), and a little more than *Bear*'s 127ft

12 MODEL SHIPWRIGHT 93

and *Terra Nova*'s 129ft. This is quite credible, as the earlier Stephen ships seem to have been quite lofty – witness the *Narwhal* model.

Judging from the photographs and confirming by more ratline-counting, it seems that the mainmast is little higher than the foremast, and that the upper masts of fore and main are identical (apart from a taller pole at the tip of the main topgallant, not present on the earlier photographs and therefore omitted from the drawing). It was usual for top- and topgallant masts, and all yards on barque-rigged whalers, to be interchangeable between fore and main, to facilitate repair.

Esquimaux's mizzen mast is interesting, being in three parts like that on *Narwhal* and other Stephen whalers which carried square mizzen top- and topgallant sails. The later barque- and barquentine-rigged ships had a two-part mizzen, with the lower mast higher than the fore to permit a large spanker to be set. *Esquimaux* is thus transitional; and one may conjecture that her mizzen was originally square-rigged, as she started life un-engined.

None of the photographs is quite clear enough to count ratlines on the mizzen lower mast, but the topmast has twenty-three and is therefore 32ft high to the crosstrees. The height of the topgallant has to be gauged by eye, as do the lengths of the jibboom, the hoisting spanker gaff and the standing monkey-gaff at the crosstrees; also the standing spencer gaff, which in photograph 1 is clearly, if unusually, set up at the rear of the main top, although later pictures show that it had by then been moved down to the level of the mainyard. As well as the ladder on the main topgallant mentioned above, there are three others each side, as drawn, taking the place of ratlines on futtock shrouds. All have wooden rungs.

On photograph A, the sails are clear enough to show cloth widths and reef points. If the twenty-four reef points on the fore upper topsail are spaced at the conventional 22in, then adding say 3ft to each side of the sail and 2ft 8in for each yard-arm gives the yard a total length of 53ft, equal to *Terra Nova*'s. I therefore accepted the dimensions of *Terra Nova*'s yards as applicable to *Esquimaux*, which was only some 18in less in the beam. However, I made the yardarms a little longer, while keeping the same overall lengths. The sixteen reef points visible on the main upper topsail and twenty-four on the foresail are more widely spaced, and a calculation based on them would make the yards clearly too short, so I stuck to the first result. As neither the painting nor photograph A shows a main course, even furled (although the plans of *Bloodhound* and *Terra Nova* do), I have not included one on the drawing.

The rigging as shown on the drawing here was derived wherever possible from the earliest photographs, ignoring later changes. Obviously, it is not possible to show or describe every belaying point or the lead of every halliard. Anyone not fully conversant with the rigging of a sailing ship of this period would do well to obtain a copy of Underhill's *Masting and Rigging*, which explains everything, from the construction of wooden spars to the correct thickness for every line and block. Only exceptions are mentioned here.

Esquimaux's standing rigging was of wire, so appears quite thin in the photographs, and should be on a model. Chain appears only on the later photographs, used for the bobstays, slings, and possibly the sheets, but was not original apart from the anchor tackle. The braces are drawn on one side only; their unusual lead keeps them up away from the funnel, the boats, and the main hatch, to give room for large bits of whale to be brought aboard. (This was done with a tackle from either the mainyard or the mainstay as the carcass lay alongside the ship, and the mainyard had to be high enough for the tackle to rotate the body while its blubber was removed.) The fore brace leads through a block attached to the first mainmast shroud; and two double blocks side by side on short pendants from the mizzen cap take the main upper and lower topsail braces. Others should be apparent from the drawing.

The barrel-shaped crow's nest, characteristic of whalers, would be hoisted to the main masthead when in use, as drawn. Access to it was by the ladder from the crosstrees and through a trap-door in the bottom of the barrel. When not hoisted, it was stowed on its side on the boat skid directly below.

Whalers usually carried eight 25ft boats, stowed on skid beams while on passage. On reaching the whaling grounds, davits were assembled and the boats hung from them. The *Narwhal* model's davits are heavy timber, mounted inside the bulwarks and supported by Y-shaped crutches outside. *Esquimaux* still had wooden davits, square section and with the top half angled outward at about 45 degrees, but without the crutches. The aftermost davit each side was of a more modern radial metal type, mounted outboard. (In later pictures, all the davits are metal radials). Two posts were fitted on the outboard side of the bulwarks between each pair of davits. The ends of adjacent boats were served by one davit, and the last two davits each side were taller, to hold the last boat a little higher. To avoid obscuring too much detail, only one boat is shown on the profile drawing. *Esquimaux* had a pair of skid beams between the mainmast and the funnel (later moved aft, straddling the funnel), and one abaft the foremast, the fore ends of the boats resting on the aft edge of the fo'c'sle, which therefore cannot have had a fixed guardrail. Photograph A shows a total of fourteen boats: in addition to the eight whaleboats, six smaller boats are stacked upside down three each side of the foremast. The picture was taken on a sealing voyage, when extra boats and crew would be carried.

As would be expected, details of the deck layout are the hardest to reconstruct. Such structures as extend above the bulwarks can be made out on photographs, but of course we have no view from above. For the rest, I have been guided by the *Narwhal* model, the deck plan of *Bloodhound* (which was, however, much altered for her naval mission), and simply by what must have been there.

The forecastle deck has the obvious items – catheads, anchors, capstan, bollards; the structure in the middle I take to be a pitched-roof skylight for the accommodation below. The vent drawn is offset to port. *Bloodhound* had ladders down to the main deck at each side, but *Esquimaux* evidently did not, and the space taken up by boats means the only place for a ladder is on the centreline. The sloping bulkhead contains two doors.

A deckhouse (presumably a galley, on account of its chimney) appears between fore and main, though only on photograph A; its panelling is conjectural. The main hatch seems right at 10ft square; the *Narwhal* model's hatch is larger, but that would require the omission of more than one deck beam. *Narwhal* also has a small booby hatch abaft the foremast, but this would have led down to accommodation which in *Esquimaux* is in the raised forecastle.

As suggested by *Narwhal*, there are fife rails each side of the mainmast, on which the braces of the fore upper yards belay, but none by the foremast. Other belaying points would be on pin rails inside the bulwarks and spider bands on each mast, but the model has none of these.

Just forward of the mainmast is a winch, and abaft it a pump (also not on the model, but they would be very likely in these positions). Next is the metal casing over the boiler room, carrying the funnel, whistle, a pair of vents, and attachment points for the mizzen stays which are duplicated each side to clear the funnel. This continues as a pitched-roof skylight over the engine-room, then a box-like structure housing a companion to the engine-room. Resting on this is the bridge, extending to the ship's sides. It has a guardrail, higher at the ends than in the middle, to which canvas screens and lifebelts would be lashed. On the bridge one would expect a binnacle and engine-room telegraph or voice-pipe to be fitted, but their positions are necessarily conjectural.

Ignoring the large deckhouse which was later built here, the area abaft the mizzen contains a skylight over the captain's cabin (conical on *Bloodhound*, but drawn here as a more conventional shape); the wheel, with gratings for the helmsmen to stand on; a pair of binnacles; an octagonal hatch over the propeller well as on *Bloodhound* (the propeller would presumably be lifted using the spanker boom); and the head of the rudder stock.

Colour scheme

Although the *Narwhal* model has green on the outer sheathing, inside the bulwarks and as a band on the boats, *Esquimaux*'s scheme contains nothing so unusual for a late nineteenth-century sailing ship. Hull, black – tarred below the waterline, 'bright' above, which on a model can be achieved by painting the whole hull matt black and satin-varnishing down to masking tape at the waterline. Not being warm-water ships, the bottom was not yellow-metalled. There was a thin white band, interrupted by the chainplates, at deck level, ie immediately below the covering board. The ship's name appeared on each side of the bow, and on the transom with the port of registry (Dundee) beneath, in white letters as required by law, but no Plimsoll lines were marked on whalers. The spurnwater round the fo'c'sle deck was white. Catheads were white. Inside the bulwarks was white, with the pinrail teak. The decks were of yellow pine or possibly bleached teak. The deck structures would probably be teak, with white panelling, and the boiler room casing white or perhaps painted brown to match the teak. The vents on it do seem to have been white, and maybe red inside. The boats were white with teak (coloured?) gunwales and thwarts. All spars were of pitch pine, finished bright, and ironwork black. The crow's nest was white, and the funnel was white with the top section black.

References

LUBBOCK, BASIL, *The Arctic Whalers*, Glasgow, Brown, Son and Ferguson, 1937

CARVEL, JOHN L, *Stephen of Linthouse*, Glasgow, Stephen, 1950

HENDERSON, DAVID S, *Fishing for the whale: a guide-catalogue to the collection of whaling relics in Dundee Museum*, Dundee Art Gallery and Museums, 1972

WEAD, FRANK, *Gales, ice and men: a biography of the steam barkentine*, Bear, London, Methuen, 1938

UNDERHILL, HAROLD, *Masting and Rigging: the Clipper Ship and Ocean Carrier*, Glasgow, Brown, Son and Ferguson, 1946 (still in print)

GREENE, W H, *The wooden walls among the ice floes*, 1933 (History of Newfoundland sealing)

FIALA, ANTHONY, *Fighting the polar ice*, London, 1907, (Account of the voyages of America)

See also the accounts of the Nares, Greely, Scott, Shackleton, Mawson and Byrd expeditions. ❑

The Aurora (1876).
Photograph: Trustees of the National Maritime Museum, Greenwich

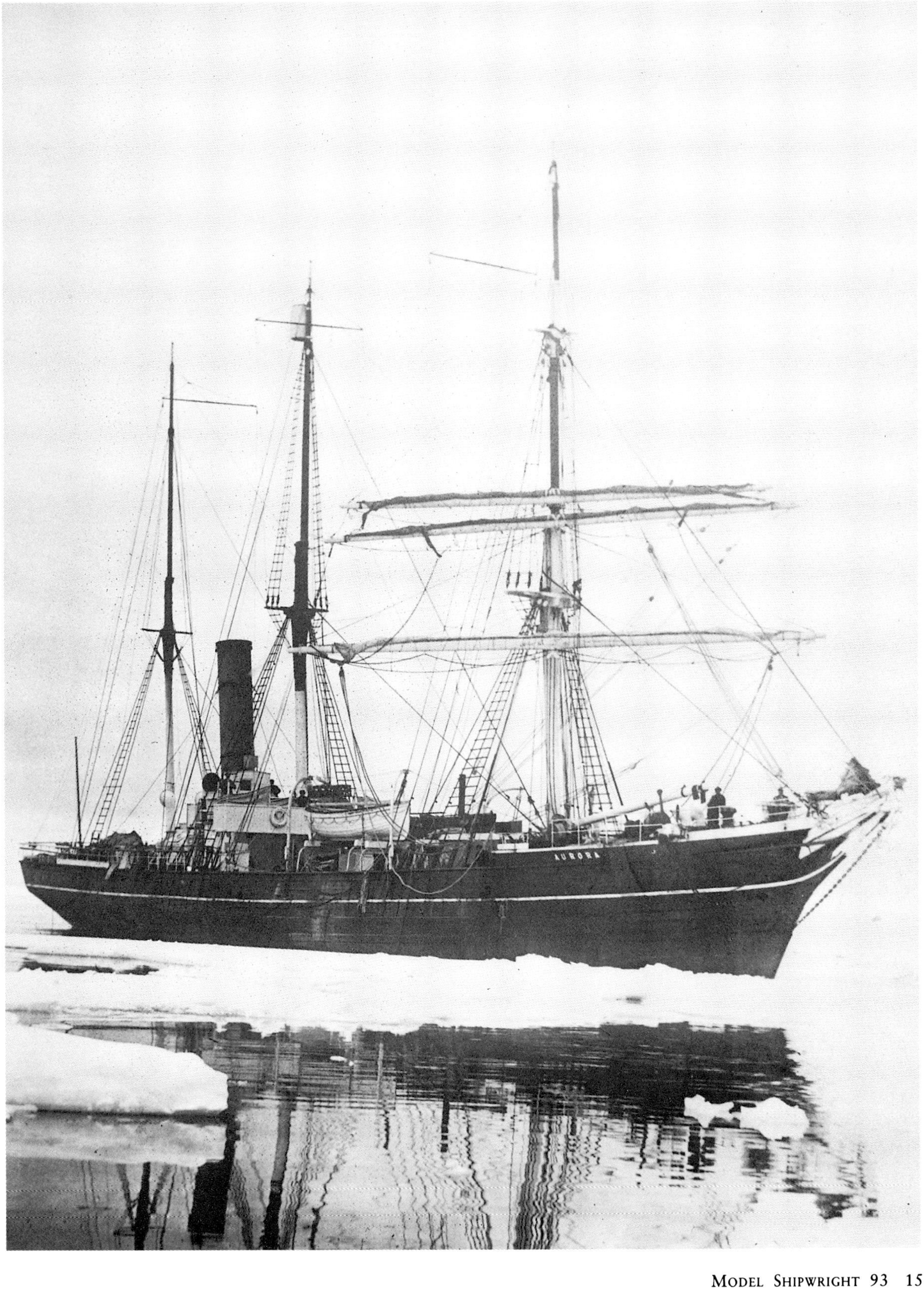
AURORA

Captain Cook's *Endeavour*

by Donald McNarry FRSA

During the extended period in the late 1970s and early '80s when I was engaged on the *Royal William* (MS 42) I decided to break up this long session by doing three waterline models of small but famous vessels much easier and less time consuming than the First rate; these were Darwin's *Beagle* (MS 69) Bligh's *Bounty* and Cook's *Endeavour*. More or less as a matter of routine I had collected a fair amount of data on all three and knew of further sources so that the research was no great problem.

The last two named are perhaps the most frequently used as subjects for models and much has already been written about them. In this journal alone there have been three articles regarding Cook's *Endeavour*, perhaps the most comprehensive of which was the earliest one by Mr R A Lightley whose beautifully-made and interesting model is now in the National Maritime Museum at Greenwich. His article (*MS* 24) covers almost all the known data sources, most notably the five original draughts at Greenwich, the further draught at the Science Museum,

Port and starboard broadside views of the finished model.

all of which show the ship at various stages of her career. Two of the Greenwich draughts have useful lists of mast and spar dimensions. Photographs are available of all of these.

There is also a set of modern plans by Harold A Underhill*. They have a few slight errors but are interesting drawings none the less. All these plans differ from each other, of course!

All the *Endeavour* models I know of are full hull, large-scale models with no attempt made to realism. I cannot recall published details of small-scale realistic or scenic models.

From this point of view one of the most helpful data sources is the book edited by J C Beaglehole, *The Journals of Captain James Cook* Vol 1, published by the Cambridge University Press in 1955. It contains numerous extracts from logs and journals kept on board during the 1768 voyage; these log entries mention quite a number of items which can be included on a realistic model but which do not appear on the draughts although one of the draughts does show the most hazardous seats of ease I have come across.

There is no doubt that the careful perusal of published logs can add much in the way of detail to a realistic model. A case in point is a most interesting and useful book on the log of the *Bounty* entitled *Mutiny! Aboard HM Armed Transport Bounty in 1789* by R M Bowker and by Lt William Bligh RN, published by Bowker & Bertram Ltd, Old Bosham, Sussex (1978). The alternative to published logs is, of course, the Public Record Office, Kew, Richmond, Surrey, TW9 4DU. If a personal visit cannot be made, then the PRO can supply a list of professional researchers who can do the work. I have had satisfactory outcomes on two occasions with this system. Photocopies of log pages can be supplied.

With regard to pictures of the *Endeavour* there are the somewhat sketchy drawings done by Sydney Parkinson who went on the voyage in 1768 (photographs are available from the British Museum) and there is a painting of the *Endeavour* in

Starboard bow view of the ship under way. The native canoe alongside is 40ft long.

Whitby Harbour reproduced in the Royal Society of Arts Journal of March 1989. This picture shows a mizzen mast of normal height; there is a school of thought that the *Endeavour*'s mizzen mast was stepped on the lower deck as shown in the sail plan in Gregory Robinson's article in the December 1935 issue of the magazine *Yachting*, but at least two of the draughts indicate that it was stepped on the keelson for the Cook voyages. The Whitby harbour picture is of the *Endeavour* when she was the *Earl of Pembroke* before being bought for the Royal Navy.

One of the Parkinson drawings is of a stern view and shows some curious square 'lids' above the four outer round top windows. When I made my model I glazed all five stern windows but I am now inclined to think that the centre one, larger than the rest and without a lid, was a dummy; inboard the panelling around the top of the sternpost and rudder would mask it considerably. On my model I omitted the lids.

For Cook's voyages the *Endeavour* was armed with ten 4pdr carriage guns and twelve bulwark or boat swivels; on my own model I have fitted the swivel guns to the stocks on the rails but have shown no carriage guns as there were two mentions in the logs of them being struck down and stowed in the hold to stop the ship rolling, or thrown overboard to lighten the ship. I was not at all averse to this omission as I had only recently made one hundred guns for the *Royal William* model.

The *Endeavour* commenced her voyage with five boats, the usual longboat, pinnace and yawl, a small

The completed model in its Australian black bean case.

Close-up amidships with native canoe alongside.

The port quarter of the model seen from above.

boat belonging to Sir Joseph Banks, the leader of the expedition, and the ship's bosun's small boat. Unfortunately, the latter was washed overboard on 1 September 1768. I have shown the pinnace and yawl nested on the booms, the longboat secured under the booms as mentioned in August 1770 and Sir Joseph's boat also on the main deck to starboard.

There is mention in the extracts of a platform over the tiller but no further description. The ship seems to have a rather long tiller so I fitted a large grating over the whole area. This is the highest part of the hull and in the interest of stability a grating would be the lightest form of platform.

The *Endeavour* was not coppered but was sheathed with wooden planks below the waterline. In June 1769 her bottom was payed with pitch and brimstone and in January 1770 she was careened and dressed with tallow and Venetian red.

There is evidence that her masts were varnished and her yards blackened and some white lead applied to her boat bottoms, but apart from this I know of no other colour information on the ship. I have assumed that she was like most other Royal Navy vessels of the time with brown varnished hull planks above black wales, blue frieze planking and yellow ochre mouldings and carving, with red edges to the gunports.

A quite detailed description of the native craft is given in the journals so I have included a 40ft native canoe for sixteen crew with 6ft paddles.

The bow of this craft is entirely taken up with a grotesque face with bulging eyes and protruding tongue. As well as a description in the logs Parkinson did a sketch of this.

The model is to a scale of 16ft to 1in, about 10in long and is shown here with her main sails aback and the native canoe moored to the starboard entry port.

The case is of Australian black bean veneer. ❑

*Harold Underhill's plans, hitherto available from Messrs Bassett-Lowke, are now being supplied by Messrs Brown, Son & Ferguson Ltd, 4/10 Darnley Street, Glasgow G41 2SD, or through some model shops. *Editor.*

Endeavour seen from the starboard quarter.
Photographs by the author

HMS *Lizard* (1697)

A 24-Gun Sixth Rate

by P S Reed

The completed model. The 3in (76mm) scale gives a good indication of the size of this 1/192 scale model.

I have long admired Donald McNarry's elegant model of HMS *Lizard*, a 24-gun Sixth rate of 1697, photographs of which can be seen in his book *Ship Models in Miniature* (David & Charles, 1975). When I found a chapter devoted to the original Navy Board model in John Franklin's book *Navy Board Ship Models 1650-1750* (Conway Maritime Press, 1989) I decided that I would have to make a model of her.

What with other projects in hand it was not until the spring of 1991 that I was able to make a start, but then with further interruptions for other models she was not completed until early 1993, after some nine hundred hours workshop work plus research time. As the illustrations show, she was built rigged and armed. The scale of the model is 1in = 16ft (1/192).

The layout of the decks.

Figurehead and bow.

I encountered comparatively few problems with research, for the original Navy Board model is in the Pitt Rivers Museum in Oxford. However, the preparatory work did take somewhat longer than I had expected, with plans to be drawn and all the necessary research for the rigging to be carried out. Being the earliest date of ship which I had modelled, I was unfamiliar with much of this.

The Pitt Rivers Museum was most helpful, and in due course and by prior arrangement I arrived at the museum to be shown into a private room, with beautiful overhead lighting and the model removed from its case and placed on a bench. I was able to spend a couple of hours taking photographs and making notes and sketches. With this information, plus a plan of the *Peregrine Falcon*, a very similar ship of the same period, I was able to prepare

The ornately decorated stern.

The finished model.

drawings for the hull. When it came to the rigging, the information came from contemporary models, from James Lees' *Masting & Rigging of English Ships of War 1625-1860* (Conway, 1979 and 1984), and R C Anderson's *17th Century Rigging* (MAP, 1955 and 1972).)

I used some holly, which I had obtained many years ago, for the frames, with the rest of the work being done in either holly or box wood. The carvings and port wreaths were carved in box wood, with artist's gesso used to add the very fine detail, before coating with Liquid Gold Leaf. The same technique was used for the dolphin supports for the model. The case and plinth were finished with myrtle veneer.

Port quarter view of the finished model; note the dolphin supports.

Looking down on the forecastle.

The quarter deck seen from above.

Mainmast details.
Photographs by the author

BOOK NEWS

Captain Cook's Endeavour; *The schooner* Bertha L Downs; *The Ottoman Steam Navy*

Captain Cook's Endeavour
By Karl Heinz Marquardt
Conway Maritime Press, 1995
240mm x 254mm (9½" x 10")
landscape, 136 pages,
25 photographs, 300 line drawings.
ISBN 0 85177 641 8. £22.00.

Captain Cook's *Endeavour* has long been a popular subject for modelmakers, and in addition a number of full size replicas have been built, including the one recently completed in Fremantle, Western Australia, so really it needs no introduction. However, in recent years it had become apparent that knowledge about the ship was not too comprehensive, and indeed some doubts were cast upon the accuracy of certain aspects of its design and outfit as hitherto interpreted and followed. More research was undertaken, and in the course of this some new information about the vessel came to light.

The author has made a deep comparative study of the three sets of surviving original draughts of the ship, and of the published and documented accounts of her voyages under Captain Cook. As a result he has formulated new ideas concerning several currently accepted points about the ship's construction, fittings and rig. He shows how he has arrived at his conclusions, which are based on his interpretation of the evidence which his researches have revealed, concerning matters such as the number of stern windows, presence of bumkins, height of the mizzen mast, stern decoration, and rig, all of which he considers up to now to have been incorrectly depicted.

The layout of the book follows the customary format of the *Anatomy* series: a short introductory text about the vessel and its history and construction, including the matters mentioned above, a selection of photographs, in this case of models and of some of the full size replicas, followed by a very comprehensive and detailed set of plans and drawings covering every aspect of the vessel, its construction, internal and external details and rig. These drawings include his interpretation of the several matters which he held currently to be incorrect. A worthwhile addition to the series, and a mine of information for all who think of building a model of this well-known vessel.

The Schooner Bertha L. Downs
By Basil Greenhill and Sam Manning
Conway Maritime Press, 1995
240mm x 254mm (9½" x 10")
landscape, 128 pages, 40 photographs,
200 line drawings.
ISBN 0 85177 615 9. Price £22.00.

This is the first volume in the well-known *Anatomy of the Ship* series to feature a twentieth-century merchant sailing vessel, and the big four-masted schooner makes an ideal subject. The large multi-masted schooners built on the east and west coasts of America, chiefly employed in that country's coastal trades, with coal as one of the principal cargoes, were fine and in the main good looking vessels, with their sweeping sheer, tall masts, and, on many, the inclusion of conspicuous, large turned wooden stanchions supporting the rails.

The book opens by tracing the history of the schooner as a type, goes on to show how it developed to suit the requirements of the American coastal trade, and concludes by describing the construction of the *Bertha L. Downs* and detailing her life history.

The photographs include a number showing vessels under construction, deck scenes, and examples of a number of other versions of the big schooners. As usual the drawings are numerous and detailed, with particular attention being paid, among others, to the ironwork for masts and spars. A very interesting note accompanies the lines plan which says that when the schooner was constructed in 1908 the hull was built from measurements (offsets) taken from a wooden half model and by established rule of thumb. These offsets were set out full size on the mould loft floor when wood patterns (moulds) were made for the ship's frames from them. Part of a table of offsets is reproduced. It was not until some years later that scale plans ie, draughts or drawings, were produced.

A useful addition to the *Anatomy* series, and one which provides modelmakers with an interesting subject for a well detailed model, whether static or working.

The Ottoman Steam Navy 1828-1923
By Bernd Langensiepen and Ahmet Guleryuz
Conway Maritime Press, 1995
295mm x 248mm (11½" x 9¾"), 192 pages, over 300 photographs and line drawings.
ISBN 0 85177 610 0. Price £35.00.

Although the history and fleets of the world's major navies of the nineteenth century have been and are well documented, that of the Ottoman Empire is not well known. Little of its technical development was known, despite Turkey's importance. The authors were fortunate in being given access to naval sources, and found that modern Turkish archives contained a surprising amount of original material relating to the period, which included many authentic plans.

The book opens with a chronological account of naval events of the period up to the demise of the Ottoman Empire and the beginning of the Turkish Republic. This is followed by a complete fleet list down to the smallest steamers, with basic technical data for each vessel, and illustrated by a large number of extremely interesting photographs, while a further bonus is the inclusion of many authentic plans, ranging from profile drawings to general arrangement plans.

The variety of vessels in service over the period is amazing, as is the number of sources from which they were obtained, either as new buildings, or secondhand. Several quite well known ships from this country and elsewhere can be found to have ended their days under the Ottoman flag. Although the same may be said of any country's fleet, time and again, as the pages are turned, there appear ships which would make excellent subjects for fine and unusual models.

The authors' researches have resulted in a book that is of immense interest and value, giving an insight into what has hitherto been something of a mystery to outsiders. Quite apart from the major contribution which it makes to naval history and as a reference book, it is one of those serious works which can also bring pleasure in a less studious way just by turning its pages – in fact by 'browsing', a pastime (often disparaged) by which knowledge of a subject is often unwittingly acquired.

MINIATURE MERCHANTMEN

ORCADES (1937)

by John Bowen

The liner *Orcades* was built by Messrs Vickers Armstrong Ltd at Barrow in Furness and handed over to her owners, the Orient Line, in July 1937, whereupon she entered the Company's UK-Australia passenger service. In the summer of 1939, with war clouds gathering, she was requisitioned for service as a troopship. After making a number of voyages in this capacity, in October 1942 she sailed from Capetown, homeward bound with a full complement of passengers, but on 10 October, just one day after leaving port, she was torpedoed. She stayed afloat and passengers and all but a skeleton crew left the ship, and were picked up by a nearby vessel. However, the following day, whilst making her way to port, she was torpedoed again and sank.

Orcades was more or less a sister ship to the Company's *Orion*, completed in 1935 at the same yard, but apart from some rearrangement of accommodation and internal layout there were several external differences, the most noticeable being that her funnel had been increased in height by 9ft, her mast shortened, and the number of deck stanchions in way of the open decks considerably reduced. She was a fine looking vessel, even if the funnel seemed to be a little on the thin side. From some angles her appearance was slightly marred by the long, almost horizontal, uptake above A deck running into the after part of the funnel.

Specification

Length overall: 665ft 0in (202.69m)
Length bp: 630ft 0in (192.02m)
Breadth mld: 82ft 0in (24.99m)
Draught (summer): 30ft 0in (9.14m)
Gross tonnage: 23,390 tons (approx)
Displacement: 28,400 tons
Machinery: Geared turbines, 24,000shp. Twin screw
Speed: 21 knots
Passengers: First 463, Tourist 605

Modelling notes

The plan is reproduced full size for a 1/1200 scale waterline model. The hull can be carved initially to the level of C deck, which includes the forecastle. The well between the forecastle and the superstructure should be cut out down to D deck level. At the after end the block will have to be cut down to the level of E deck and again at the after end of that deck to the level of the top of F deck bulwark. With that done the hull can be carved to shape. Once the forward and after ends of the well have been finished, the bulwarks each side fitted, and the whole area painted, the small deckhouse therein can be fixed in place. The forecastle will have to be covered with a deck card, extended to cover this house, and shaped as shown on the plan. At the stern a pencil line must be drawn round, and about 1mm in from the edge of F deck. The surplus material inside this line must be cut away out to the line and down to a depth of 3ft 6in, to form the bulwark round the stern. The 1mm wide bulwark can be pared away to leave a narrow top edge.

MODEL SHIPWRIGHT 93 27

28 MODEL SHIPWRIGHT 93

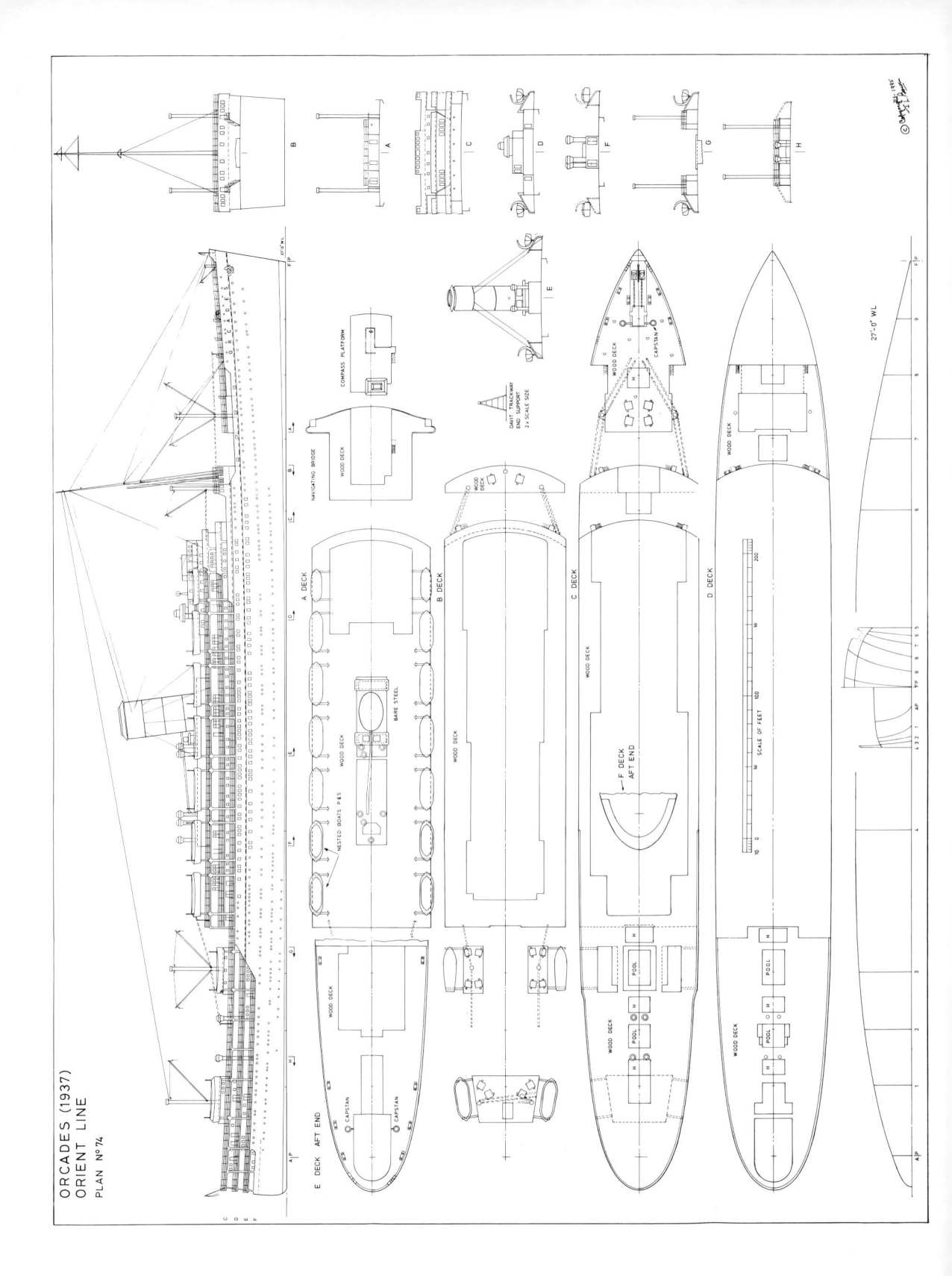

Orcades.
Photo: Orient Line

After adding the shaped, short end of the deckhouse the area must be painted. The after end of the hull should be built up, with finished deckhouses, hatches, etc and deck cards to C deck level. The card for C deck must be cut to cover the whole of the deck from the fore end of the superstructure to the stern, and include the 2ft wide overhang each side. This overhang ends at the fore end of the open side of D deck aft.

The remaining houses and deck cards forming the superstructure are straightforward, but the deck cards must be cut to match the overhang of C deck. There are two swimming pools on C deck and holes for them will have to be cut in the card and the houses on D deck below before these are fitted in place. Each pool has a shallow surround, cut from two-ply bristol board or similar card.

The lifeboats are carried on gravity davits, the inboard end of the majority of the trackways being supported by A-frame units. The two boats each side at the after end of D deck, and the boat each side at the after end of C deck each have a second boat nested in them. The other boat each side of C deck is a motor launch.

The funnel is oval, and care must be taken when making the cowl top, as its shape is an important feature of Orient Line ships. Two cylindrical water tanks are fitted on top of the large rectangular vent immediately abaft the funnel, with the large uptake mentioned earlier passing between them.

Colour scheme

Hull: the colour of the main hull is invariably described as 'corn colour'. The nearest approximation to this is about four parts Humbrol 7 with one part Humbrol 24. The top edge of this hull colour is level with the bottom edge of the curtain plate to D deck aft, and follows the line of the sheer to the bow. Above this line the hull is white. Boot topping green.
Superstructure: white, deckhouses white, inside bulwarks white.
Masts, derrick posts, derricks: light buff, Humbrol 24.
Ventilators: large conical ventilators on engine and boiler casing funnel colour, elsewhere white, inside cowls buff.
Lifeboats, davits: white, covers green.
Windlass, winches, capstans: light buff.
Bollards, fairleads: buff.
Hatches: light buff.
Funnel: light buff, cowl top black.
Decks: planked, areas of bare steel deck black.
Miscellaneous: bulwark on Compass Platform varnished teak. Anchors black. Anchor crane white. Long uptake on A deck funnel colour.

References

Shipbuilder & Marine Engine-Builder issues for August and September 1937 contain an illustrated article with small scale general arrangement plans.

Shipbuilding & Shipping Record issue for 26 August 1937 has a similar article.

For general reading about the Orient Line: *Passenger Ships of the Orient Line* by Neil McCart, published by Patrick Stephens (1987).

Note: *Orion* was featured in the September and October 1935 issues of *The Shipbuilder & Marine Engine-Builder*.

1853 Whaling Schooner

Notes on a scenic model

by Peter Heriz-Smith

The New England whaling schooners were not uncommon in northern waters; but the only ones portrayed in contemporary prints are the square riggers, as are the models I have been able to see in the Science Museum, the Sydney Museum and the American Museum in Bath. The only survivor, the *Charles W Morgan*, which after eighty years service rests in her final berth at Mystic, Connecticut, is also square rigged.

A model of a schooner, therefore, offers an interesting change. It was V R Grimwood's book *American Ship Models* that first attracted me to the idea of making one. He describes in some detail the *Agate*, built in Massachusetts in 1852, and gives plans based on lines taken from the builder's half-model by Howard Chapelle.

Time passed, during which I took a casual and intermittent interest in the subject of whalers and noted some possibly useful references; but it was not until recently that I decided that a scenic miniature showing every aspect of the nineteenth-century whaling business would be of interest and enjoyable to build.

Agate was a small ship, at 79ft 10in O/A some 30ft shorter than the average square-rigger. She was sturdily built with considerable carrying capacity, for she and her crew had to be completely self-contained for sometimes years at a time before returning to their home port of Provincetown with, it would be hoped, a capacity cargo of whale oil. Unlike the Gloucester fishing schooners, she had no need of speed.

Although she carried a normal schooner rig several features unique to whalers could be seen. She was, in whaler classification, a three-boat vessel, and the graceful whaleboats carried at port and starboard davits were distinctive. The third lay under stern davits, and was probably held in reserve. These whaleboats were very lightly built and weighed only about 1000lbs. They could be – and often were – smashed to matchwood by a wounded whale 'breaching' – hurtling from the depths to and sometimes out of the

The completed model of the diorama.

water. Because they were working boats, they were carried outboard for ease of launching. The davits were distinctive. Usually, they were curved, but sometimes they were built in a sort of gallows, and Grimwood shows the latter in his plans for *Agate*. Since I was not 'naming' my schooner, I felt free to adopt the more elegant curved version.

Because of the light construction of the whaleboats and the weight of specialised gear they had to carry, suspension from davits would put too great a strain amidships. Subsidiary support was therefore given by a couple of brackets, hinged so that they could be retracted against the ship when the time came for launching. The sketch (Figure 2) illustrates these points.

[Construction of the whaleboats was described in detail in the article on a New Bedford whaleboat by Pat Earle in *MS* 92, and is also discussed in the article on the *Agate* by Peter Rogers in the same issue. Editor].

As far as the inboard arrangements and fittings are concerned, these cannot be isolated from their purpose, so I will not try to bring them into context – essential if a scenic model is intended. What follows, then, is a very much abbreviated account of what went on in the course of a typical day; and it is hoped that the accompanying drawings will help to clarify the text.

1. At the masthead cry of 'Thar she blows!', the two operational boats are launched, and the chase begins. The helmsman with his long oar steers close enough to enable the harpooner to thrust (not to throw) his harpoon or iron into the whale.

2. What is called a 'flurry' ensues, in which the whale 'sounds' in a deep dive. The line to which the harpoon is attached snakes out of the two tubs in which it is coiled, aft round the loggerhead and out through the fairlead in the bows. The line round the loggerhead smokes from the friction and has to be cooled with water.

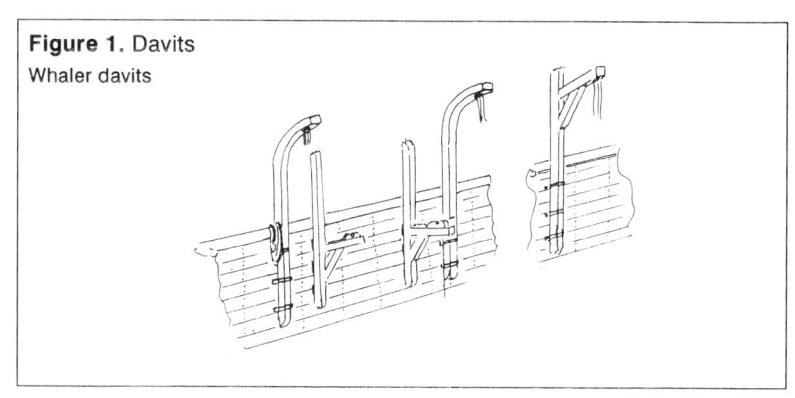

3. When the wounded whale runs out of air, it hurtles to the surface – called 'breeching' – and takes off with the boat in tow on a 'Nantucket Sleigh ride'.

4. Eventually, the exhausted whale can be killed with a special lance, and towed back the mother ship.

5. In the meantime, on board the whaling ship the cutting-in stage has been lowered, and the crew stand by to receive the whale for flensing (Figure 2). The carcass is first secured, fluke forward, by the fluke tackle led through a hawse in the bow. It must now be brought into position alongside and under the cutting-in stage. However, the whale's inertia in the water would tend to pull the bow round to starboard and it would float free at a wide angle to the hull.

6. To counter this, the ship had to be hove-to in order to achieve a sort of controlled forward drift with just enough way to enable the rudder to 'bite'. In the case of a square-rigger, the method is well described by Harland (see Bibliography) and amply illustrated in contemporary prints. Briefly, forward way was maintained by the mainsails, and checked by backing the lower foresail and lowering its topsail to mask it. The forward fore-and-aft sails were sent down, and the helm laid to starboard. In this way, the whale and hull were nudged into contact.

7. What happened in a schooner is less clear. Harland suggests that normally a schooner could be hove-to by backing the fore main. This would certainly not work in a whaler, for the boom would become tangled up with the cutting-in tackle. In my model I have imagined a very light breeze and the schooner

Figure 3. The whale
1. Lead taken to windlass via masthead tackle
2. Blubber hook
3. Fluke tackle
4. Fluke
5. Blanket. The whale revolves as the blanket is cut away and lifted
6. Jaw – valued as a source of material for scrimshaw.
7. Case
8. Source of pure spermaceti oil

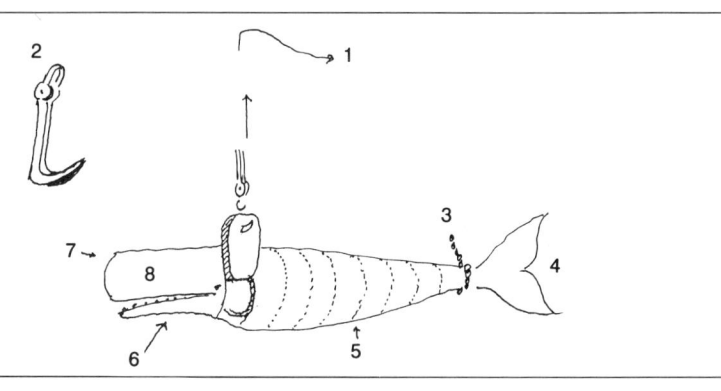

Figure 4. Try works
1. Try pots
2. Fireboxes
3. Cooling tank
4. Skimmer
5. Bailer
6. Blubber fork
7. Mincing horse board
8. Horse pieces
9. Slicing into 'Bible leaves'
10. Mincing knife

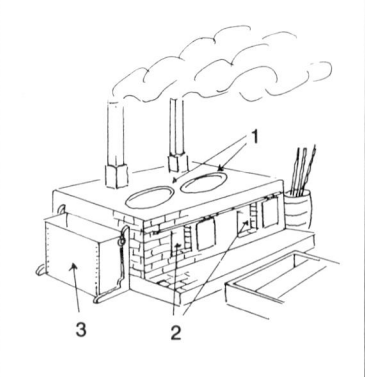

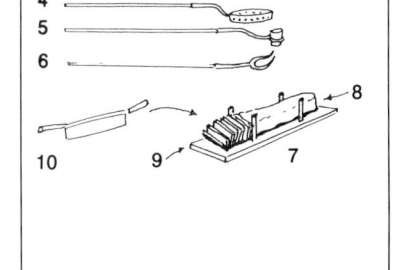

Port bow view showing try works and davit details.

reefed right down. An alternative in heavier weather might be to back the topmast foresail (the 'fisherman').

8. Flensing can now go ahead. One of the crew climbs down onto the whale, secured by a lifeline, and inserts a blubber hook into the blubber behind the eye. The crew on the stage get to work with their cutting spades, and as they free the blubber from the carcase in an endless spiral, the 'blanket' is torn upwards by the cutting-in tackle, while the whale slowly turns on its own axis. Every 15ft, a section of the blanket is cut off and lowered through the main hatch to the blubber room; and the process is repeated until the blanket is completely removed (Figure 3).

9. In the blubber room, each section of blanket is cut, first into 'horse pieces' and then into 'Bible leaves' (Figure 4) in which form it is taken on deck and fed into the try-pots of the tryworks; the oil is boiled out of them and ladled into the cooling tank before being transferred into casks and stored in the hold. The remnants of the Bible leaves are skimmed out and fed into the fireboxes, and it is this which produces the clouds of black oily smoke typical of these craft. It is an interesting recycling process by which the dead whale effectively fuels its own destruction. I hope that the sketches help to make the whole thing comprehensible.

Figure 5 is an outboard and inboard elevation and deck plan of the whaling schooner, developed from the plan in Grimwood's book. The dimensions of *Agate* shown below were taken from that work.

Specification

Length: 79ft 10in
Length on load waterline: 74ft 0in
Beam moulded: 19ft 8½in
Beam, extreme: 20ft 1in
Depth in hold: 8ft 6in
Draught at post: 10ft 0in
Room and space: 2ft 0in
Bowsprit: 23ft 9in
Bowsprit, outboard: 15ft 0in
Jibboom: 28ft 0in
Jibboom, outboard: 15ft 0in
Foremast to cap: 60ft 0in
Fore topmast: 40ft 0in
Fore gaff: 24ft 0in
Fore boom: 24ft 0in
Mainmast: 64ft 0in
Main topmast: 41ft 0in
Main gaff: 29ft 0in
Main boom: 49ft 0in

Hull

The method of building a basic hull structure (Figure 6) from laminated blocks is orthodox, and has been covered before. Only two points need clarification. Lamination A has

The 'flurry'.

This, then, describes what I have tried to illustrate in my 1/192 scale model. But what cannot be reproduced is the stench of blood, blubber and acrid smoke which surrounded whalers in a foul miasma, so that they were known to other seamen as 'slaughterhouses' and 'stinkers'.

The model

These preparatory notes may seem to be a bit excessive, but they serve to emphasise the importance of at least a superficial understanding of the way a vessel works before construction starts. This is especially necessary when the prototype is unfamiliar and highly specialised, and the object is to show it in action.

Figure 5. General arrangement
Inboard profile

Deck plan

Starboard side profile

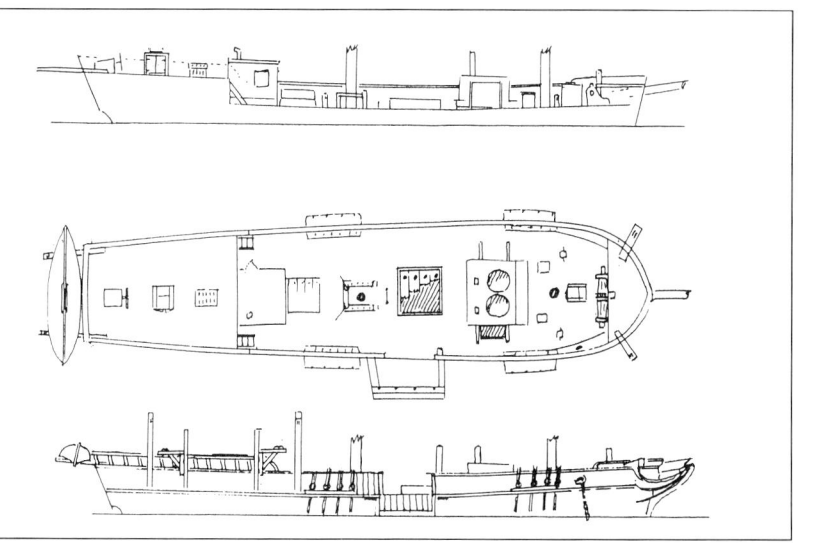

Figure 6. Hull construction

Line of sight
Main hatch
Forecastle

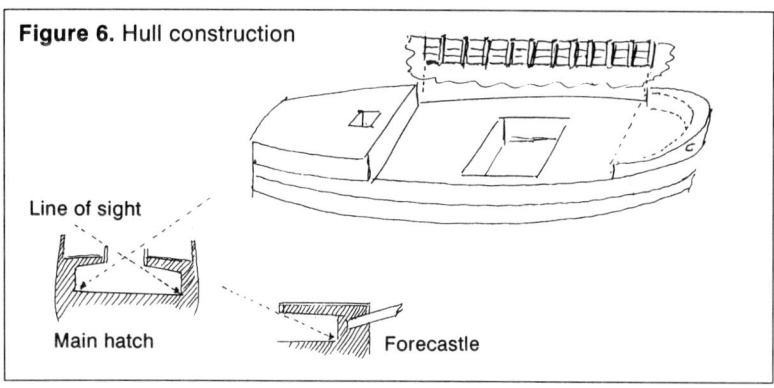

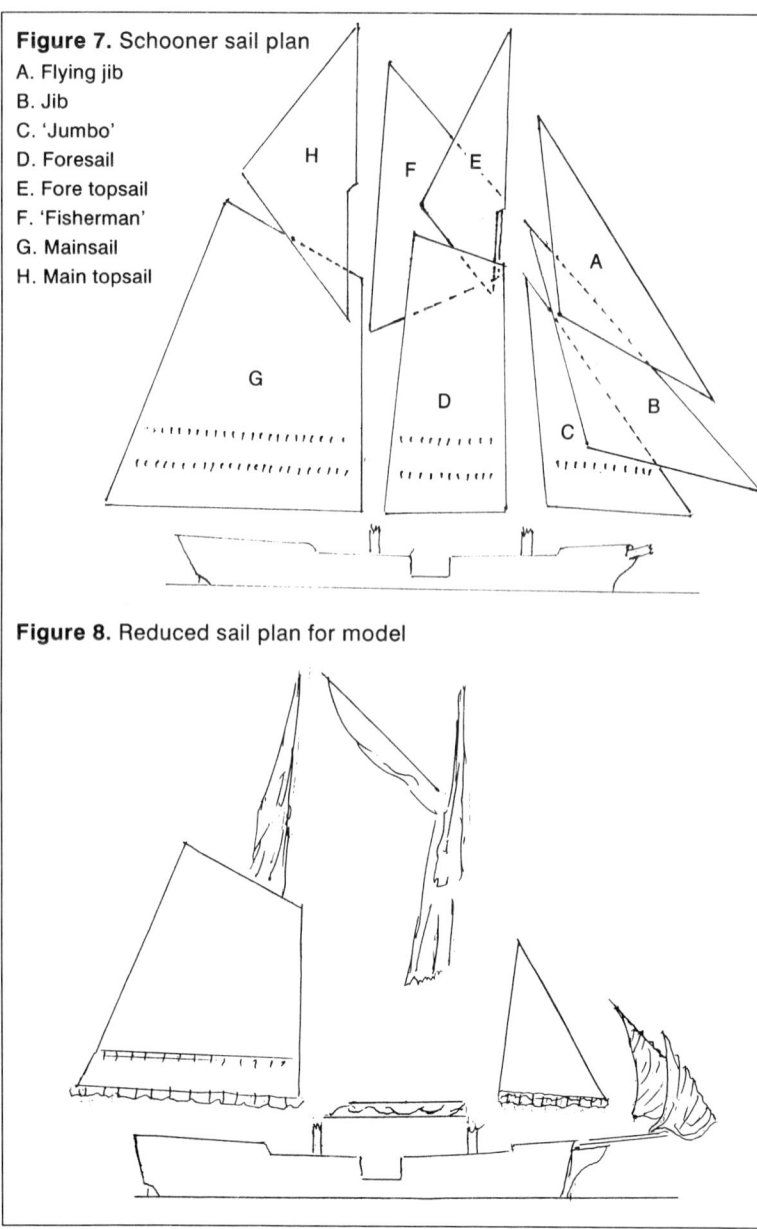

Figure 7. Schooner sail plan
A. Flying jib
B. Jib
C. 'Jumbo'
D. Foresail
E. Fore topsail
F. 'Fisherman'
G. Mainsail
H. Main topsail

Figure 8. Reduced sail plan for model

a large opening cut out in way of the main hatch, which is shown open. This is in accordance with the 'line of sight' principle, which relies on the illusion that a space which extends as far as the eye can see must continue beyond that point. This useful principle applies to many aspects of miniature work.

The laminations can be permanently glued together except for the forecastle block C. This need only be spot glued centrally as it will have to be removed later to be hollowed out as shown by the dotted line. The hull can now be shaped and carved in accordance with the customary templates, and the decks added. These are veneer planks laid on good quality paper and trimmed to fit. I must admit here that I ignored the camber of the decks which are so cluttered that it would have been undetectable.

The bulwarks are constructed separately on oversize paper bases which can be faired into the hull, after which the whole of the outside of the hull can be planked. Self-adhesive copper tape was used for coppering below the waterline. The addition of wales, rails and bulwark capping presented no problem. The gap in the starboard bulwark in way of the cutting-in stage should be left to last, in order to preserve the line of the hull. If cut out earlier distortion might occur.

Try works

These are shown in Figure 4. Although this unit is only $5/8$in square, it is such an important feature of any whaler that it demands particular attention to detail. It is constructed over a solid block which has been drilled and recessed in way of the try pots and fire boxes. The prototype was built of brick, and this must now be indicated.

A sheet of thin paper (typing paper will do) is covered with fairly fine lines drawn with watercolour pens or a fine brush in varying shades of 'brick'. The paper is then turned through 90 degrees and sliced into thin strips to represent the courses. These are now glued to another bit of paper, each being staggered so that the vertical lines of colour are lost. The try works block is veneered with sections of this brickwork, and the end result is surprisingly convincing.

The rims of the trypots are sawn from aluminium tube of the correct diameter, blackened and fixed in place with superglue. The surface of the oil content is a disc of transparent plastic. The cooling tank is shown half full. The fireboxes are painted inside with red and yellow, with a few twisted scraps of tissue, painted orange and black, representing the burning material. Chimneys are of aluminium tubing, and smoke is imitated with wisps of cotton wool. This material cannot, of course, be painted, but grey pastel does the job effectively.

Cutting-in stage

This is a prominent feature (Figure 2). It is easy to make, but very delicate. To fix it rigidly to the hull, the point of attachment was sharpened, and pushed into slots below the wale opened with the point of a knife. Touches of superglue completed the job.

Other deck furniture

The galley and one of the companionways are shown with open doors.

Sails

The full suit of sails for a schooner is show diagrammatically (Figure 7). As explained earlier, my whaler is ghosting along under the mainsail and the 'Jumbo' – both reefed. The foresail, jib and flying jib are sent down, and the topsails and 'Fisherman' hang loose. They might, however, need to be brought into play with a change of wind or weather (Figure 8).

Furled sails at this scale are a bit of a problem, for they cannot be worried or manipulated into convincing folds. A method I have found to be effective is illustrated in Figure 9. A paper silhouette of the sail is cut out, and thin twists of tissue paper are rolled between gluey fingers and applied to both sides of the blank to indicate folds in the material. When the sail is trimmed the method is very difficult to detect.

Blocks & deadeyes

Various methods of making these in miniature have been described in *Model Shipwright*. I mass-produce mine with a material called *Fimo*. This is a modelling paste which can be hardened in the oven. It is rolled into 'sticks' of various suitable diameters, and these, when cooked, can be sliced into segments as required. If the resulting block looks too circular, the fault can be cured by slicing at a slight angle.

Figure 9. Furled sails

Bringing in a whale.

Whaleboats

These are made from paper laminated over a solid core and fitted out with correct detail – much of which will be obscured by the seven-man crews. I show two in the water. The third on the stern davits would have been secured keel downward to show the details more clearly. However, after I had dropped it and trodden on it I had not the heart to make another; hence the boat shown secured upside down.

Whales

Since the three shown are all waterline, construction is very simple, and they are carved from lime. The blanket piece being lifted by the try tackle is a piece of mounting board, as is the one being lifted into the blubber room. Being wet, the whales are gloss varnished.

Crew

There are thirty-five of them, all usefully employed in some task or other. They are built up with glue and acrylic paint over wire armatures. As usual, I have relied on action, proportion and attitude rather than attempt the (for me) impossible task of introducing detail. Those carrying or wielding implements or oars have these things spot glued to the end of their arms with superglue, and their hands are then added with acrylic.

The sea

There is only a light wind and a calm sea in my model, so there is no drama. A 16in x 7in piece of mounting board was used and recesses cut in it to take the whaler and its two boats. It was then strongly battened underneath and the water surface was indicated with filler, using my fingers. I find it impossible to describe this process, but I tried with restless finger-movements to reflect the restlessness of the sea, smearing and dabbing and adding where necessary. Acrylic paint in shades of blue, brown, ochre and yellow were smeared on, again using my fingers. A touch of blood was indicated in the wake of the whale being flensed, and the whole finished off with several coats of varnish. The underside was given a protective coat of paint, and a frame of hockeystick section added to round things off.

Bits and pieces

On board the whaler there remained a few minor items to be added to the clutter – spare flensers' spades, rope ends and rope coils, a grindstone, barrels, a whale's jaw and the equipment for the spare boat. Outboard, I have shown gulls attacking the flensed carcass, and a couple of sharks approaching for their share.

Although the actual scene is a trifle condensed, I have given the model enough sea and air to indicate a very small ship drifting without destination in a very wide ocean, and the crews going about their demanding and sometimes dangerous duties. With all the mass of congested details involved, the construction gave me many hours of pleasure.

The model was awarded a Highly Commended Certificate at the 1995 Model Engineer & Engineering Exhibition in London.

References

V R GRIMWOOD *American Ship Models*, Bonanza Books, New York. 1942

A B C WHIPPLE *The Whalers*, Time Life Books, Amsterdam. 1979

JOHN HARLAND *Seamanship in the Age of Sail*, Conway Maritime Press. 1984

E J MARCH *The Gloucester Fishing Schooners, Ships & Ship Models*, issues for June 1956 & seq for schooner rig.

CHARLES G DAVIS *Ships of the Past*, Bonanza Books, New York. 1929

H C HOLLING *Seabird*, Collins. 1960. Although this is written for young readers, it is well researched and accurately illustrated in line and colour and I found it inspirational in the work on this scenic model.

Flensing: note the 'blanket' being lifted by the masthead tackle.

Photographs by the author

Anchor Work in the Royal Navy

Part III: Iron ships, steam reciprocating engines, chain cable & close stowing anchors

by David White

The use of iron and steel in shipbuilding in the latter part of the nineteenth century resulted in ships becoming progressively larger and consequently requiring heavier ground tackle. Since the early part of the century, when wooden stocked anchors with hempen cables had been the norm, chain cable had been introduced and many anchors had acquired heavy iron stocks (see Parts I and II of this series). However, because chain cable was much heavier than its hemp equivalent, the actual weight of anchor for any given size of ship was actually less. For example, the bower anchors of the 120-gun *Caledonia* of 1808 weighed, with stocks, 5.85 tons and required 25in hemp cables, the weight of which for 100 fathoms was 5.78 tons. The equivalent chain cable was 2¼in and 100 fathoms weighed 12.15 tons or 2.1 times as much as the hemp. Consequently, the iron steam ship *Temeraire* of 1876 with a load displacement of 8415 tons, which was slightly more than twice the *Caledonia*'s of 4158 tons, was fitted with iron stocked bower anchors of only 6 tons.

The introduction, around 1870, of steam-powered capstans and cable holders had made the handling of cables easier but the catting and stowing of these large anchors using traditional methods was a cumbersome affair, (the shank of a 6-ton anchor was over 22ft long and the stock was over 24ft). Furthermore, with the introduction of upper deck guns the catheads and the stocks of the stowed anchors frequently fouled their arcs of fire. The solution was found in the invention, by Francois Martin in 1859, of the close stowing anchor (Figure 1). After a decade of modifications and testing in Woolwich dockyard it was approved by the Admiralty for service in the Navy.

In this type of anchor the stock was approximately of the same length and in the same plane as the arms. The arms were not fixed to the shank but pivoted on a pin or ball through the crown. Stops were provided to limit the angle through which the arms could turn which was usually about 40-45 degrees. When it was drawn along the seabed the arms were tripped by large flat 'tripping palms' at right angles to them, causing the flukes to dig in and hold the ship. There were several makes of close-stowing anchors, differing basically in the shapes of their palms and the methods of hinging them. The Royal Navy mainly used the Martin pattern but other makes including Inglefield (Figure 4), Hall, Brown-Lennox and Baxter were also employed.

The usual way of stowing this type of anchor was on an 'anchor bed', known initially by the old sailing ship name of 'billboard', which was a flat plate let into the forecastle deck, sloping outwards at about 20 degrees to the horizontal, with its outboard edge rounded down and with a recess at its after edge to

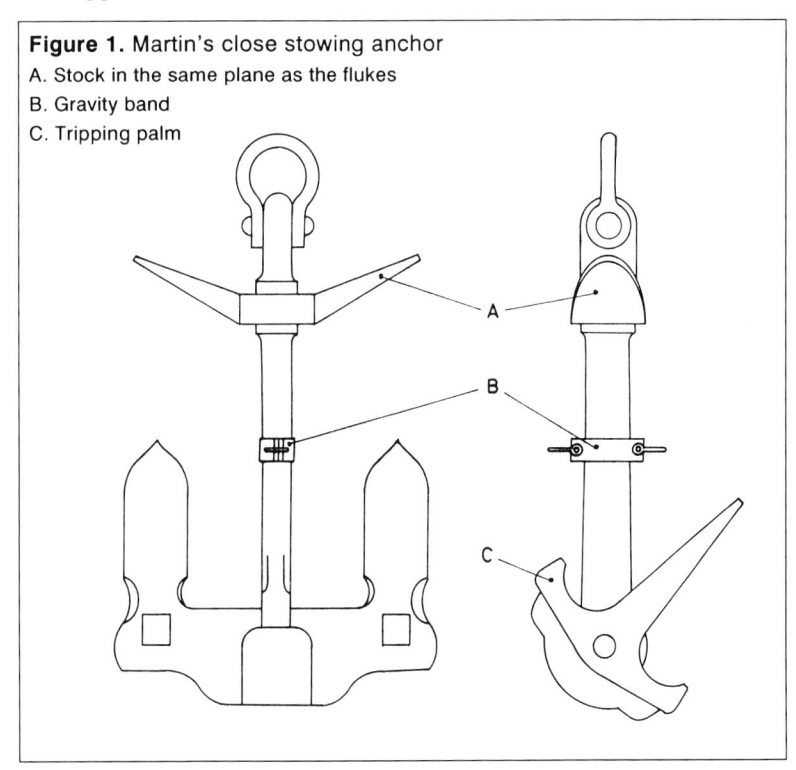

Figure 1. Martin's close stowing anchor
A. Stock in the same plane as the flukes
B. Gravity band
C. Tripping palm

Figure 2a. Plan of anchor bed
A. Cat davit
B. Release tumbler
C. Skids
D. Recess for tripping palms
E. Bed
F. Tumbler chains, known originally as shank painters
G. Seizings or bottle screws
H. Cat davit guys

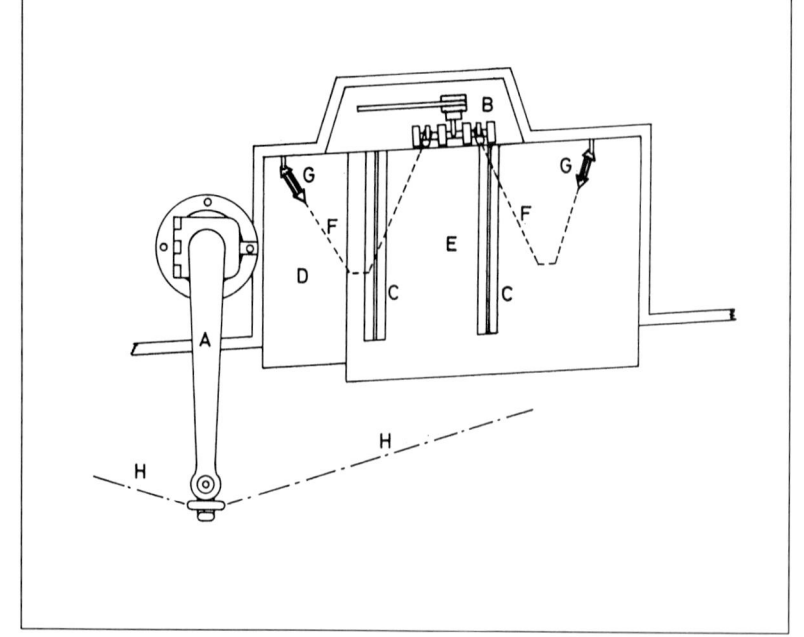

Figure 2b. Profile of anchor bed
A. Cat davit
B. Cat pendant
C. Release tumbler
D. Recess for tripping palms
E. Skids
F. Bed

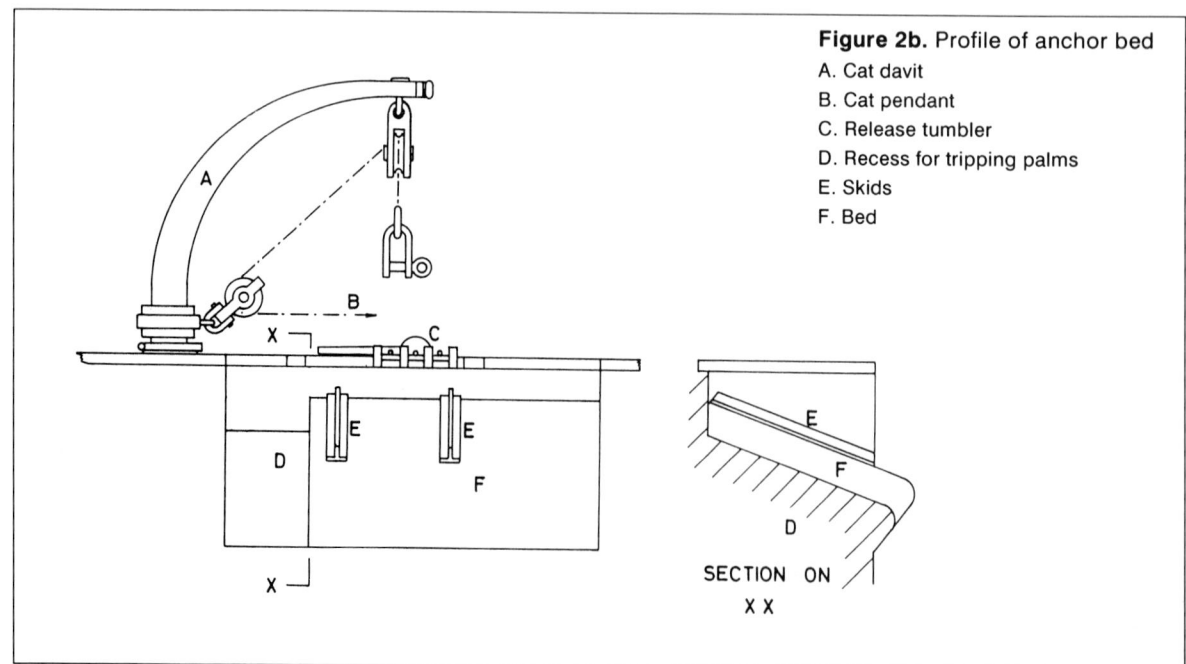

SECTION ON XX

accommodate the tripping palms (Figures 2a & 2b). Each bed was fitted with two skids on which the anchor rested and on which it slid when released. Early destroyers with turtle-back forecastles had no need of an anchor bed, the skids being attached directly to the forecastle deck.

Two 'tumbler chains' (known initially by the old sailing ship name of 'shank painters') were used to secure the anchor. Their standing ends were either fitted with triangular links which were seized to similar links attached to the bed, or with bottle screws shackled to eye bolts on the bed. After passing round the shank of the anchor, rings in the running ends of the chains fitted over two horns on a 'tumbler bolt' (Figure 3). Between the horns the bolt had a 'spill' which pointed inboard and which was held in the stowed position by a semi-circular tumbler attached to a release lever. To let go the anchor the lever was raised, allowing the bolt to turn and release the ends of the chains from the horns. The anchor was then free to slide down the skids, assisted if required by the judicious use of a handspike.

When securing for sea extra chains were added, one through the anchor ring, one round the inboard arm and two round the shank. The standing ends of these chains were fastened to eye-plates on the bed in the same way that the shank painters were. The other ends had triangular links which were seized to similar links on the bed; alternatively, they were secured with bottle screws and slips.

Although strictly beyond the scope of this article, it is interesting to note that when the 'P' Class Patrol Boats were first built in 1916 it was found that their stockless anchors, being very close to the waterline because of their very low freeboard, threw up unacceptable amounts of spray over the forward gun mounting and bridge. This was rectified by fitting them with close stowing anchors on anchor beds further aft and closing their hawsepipes with bucklers.

The 1st Class cruisers *Powerful* and *Terrible* of 1895 and the *Diadem* class of 1896-98, together with the armoured cruisers of the *Cressy* class of 1899-1901 and the early members of the *Monmouth* class of

Figure 3. Release tumbler for a close stowing anchor on an anchor bed
A. Securing pin
B. Release lever
C. Tumbler
D. Spill
E. Horn
F. Tumbler chain
G. Tumbler bolt

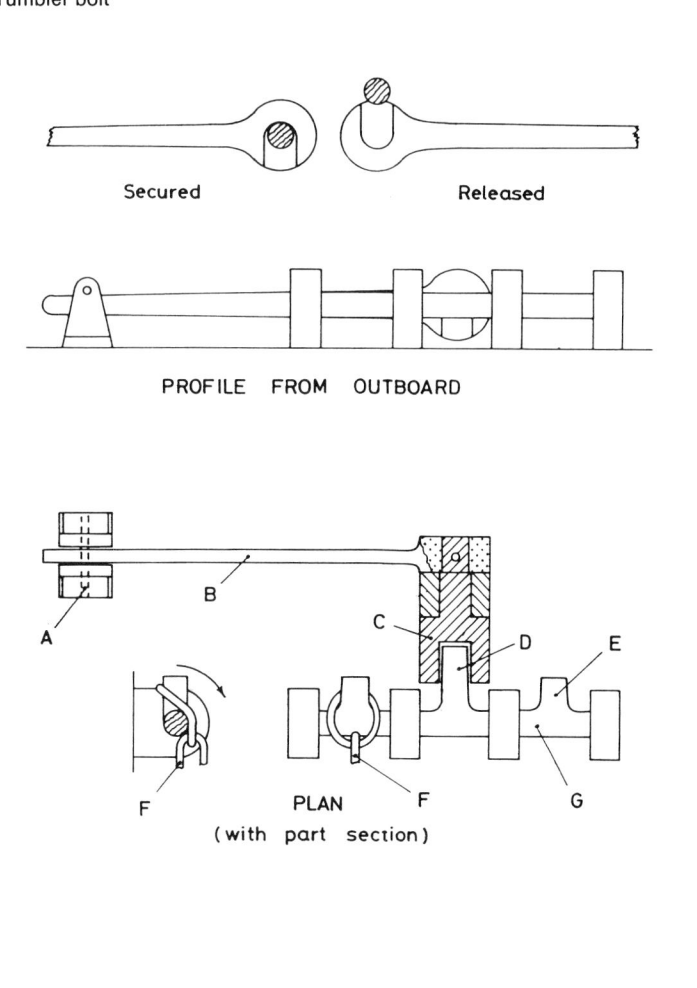

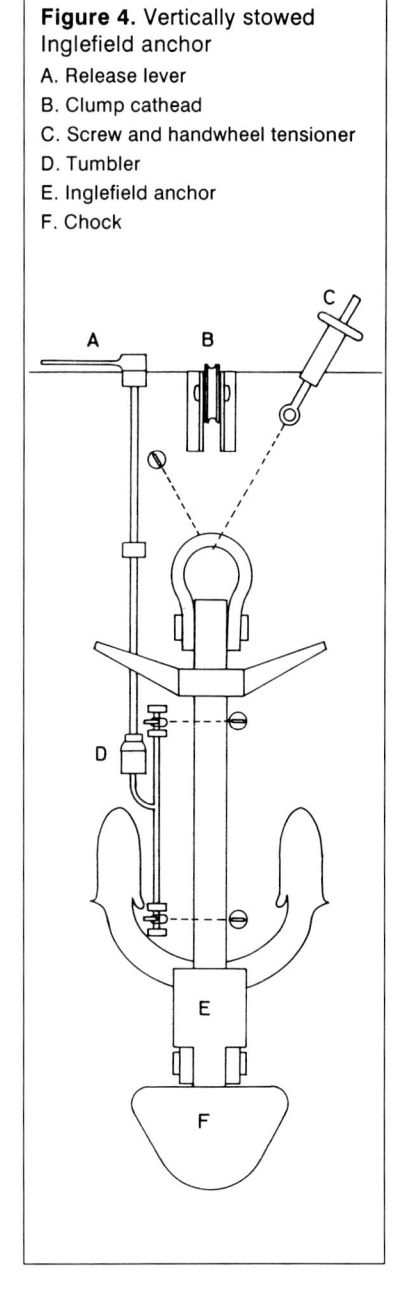

Figure 4. Vertically stowed Inglefield anchor
A. Release lever
B. Clump cathead
C. Screw and handwheel tensioner
D. Tumbler
E. Inglefield anchor
F. Chock

Figure 5. Blake slip.

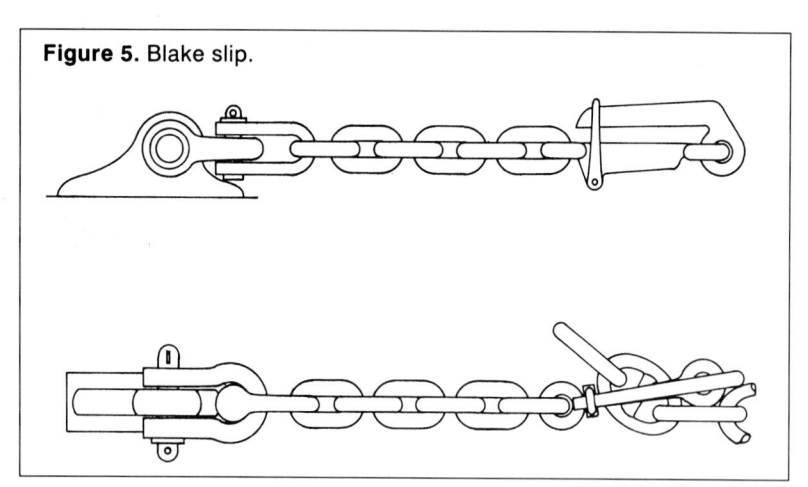

Figure 6. Senhouse slip
A. Deck pipe
B. Slip
C. Top of chain locker
D. Deck clench

1901 all had very high freeboard and were, consequently, able to stow their anchors vertically, below the forecastle deck edge (Figure 4).

This had several advantages over stowing on a bed, not the least of which was the fact that the chock which replaced the bed was simpler and therefore cheaper. Its main advantages, however, were first, that the rather restricted forecastles in cruisers were not further encumbered with anchor beds and second, that the anchors no longer needed 'ground chains' fitted in order to weigh them. (This will be explained later under the heading 'Working cable – weighing'.)

Vertical stowage of close stowing anchors was not taken any further because the introduction of the stockless anchor in 1902, with its even simpler method of stowage in the hawsepipe, rendered it obsolete.

This article only deals with the use of close stowing anchors as bower anchors. They continued in service for many years as stream anchors and destroyers were still being built with them as bower anchors up until 1909.

Cables

There was very little difference in the wrought iron chain cable used in this period from that of the previous one. (see Part II of this series) Experience had shown that it was not necessary to have a swivel in each length, or shackle as it was now known. Consequently, only the first and last shackles had one, in the outboard and inboard ends respectively. The joining shackles were still marked by wire around the studs, but now they were marked on either side of the joining shackles, on the first studs for the first joining shackle, the second studs for the second and so on.

The slip stoppers which had taken the places of the bitt stoppers were replaced with a modified version, the 'Blake slip stopper' (Figure 5) and the slip stopper in the chain locker was replaced by the 'Senhouse slip' (Figure 6). This differed from the other slips in that the tongue went through the last link of the cable instead of going across it.

Capstans and cable holders

The introduction of the steam-powered capstan, in the late 1860s, reduced the manpower required for working cables. This was not such a great advantage as it may seem. As the ships of the day still relied heavily on sail power they still had large numbers of seamen to handle the sails, many of whom were available, together with the marines, to man the capstans when weighing anchor. The initial advantage of the steam capstan was its ability to speed up the process.

Originally, capstans were made to heave on hemp cordage and cables were worked using a hemp messenger. When chain cable was first introduced, in 1811, it was still worked using a messenger but by 1870 steam capstans were in use that could work chain direct as well as hemp cordage (Figure 7). The barrel was equipped with cast iron 'snugs' to grip the cable and port-

Figure 7. Harfield's capstan
A. Capstan bar shoe
B. Securing pin for capstan bar and/or portable whelp
C. Key
D. Barrel for working small steel wire or hempen rope
E. Snugs for cable
F. Pawl rest
G. Portable whelp for working large steel wire or hempen hawsers
H. Pawl
I. Pawl rack
J. Pawl bed
K. Spindle

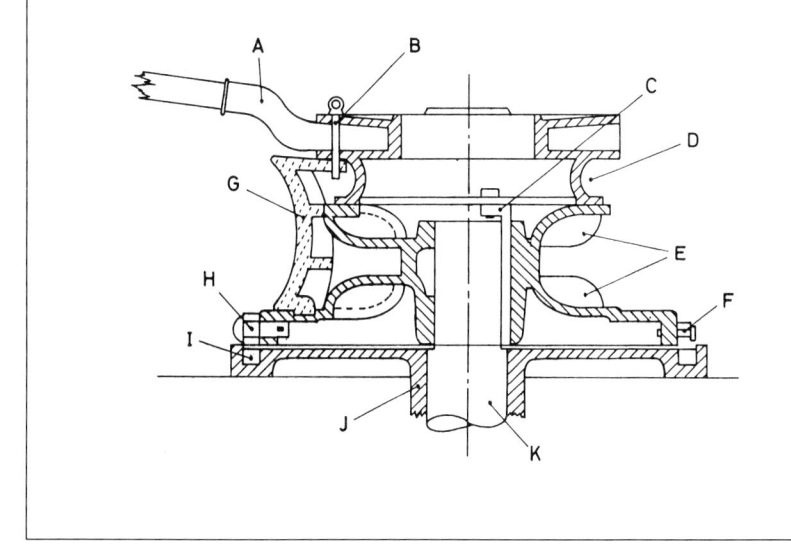

able whelps were supplied for use with hawsers. Capstans were reversible in that they could heave in either way but what they could not do was veer cable by 'surging' it.

The next move was to introduce a capstan to work chain cable only, but which could veer cable under control and 'bitt' it too. This became known as a 'cable holder' (Figure 8). By providing one for each bower cable, independent of the capstan and each other, the capstan was left free for other uses, but remained available for cable handling in the event of a breakdown in either one or both of the cable holders. Unlike the steam capstans, which could be worked by hand in the traditional way, cable holders had no manual facility. The advantage of being able to work cable in different directions on opposite sides at the same time, without having to change over messengers or reverse capstans, was enormous, drastically cutting down the times taken for manoeuvres such as mooring.

When weighing with the original Harfield and Clarke-Chapman cable holders, the drum was connected

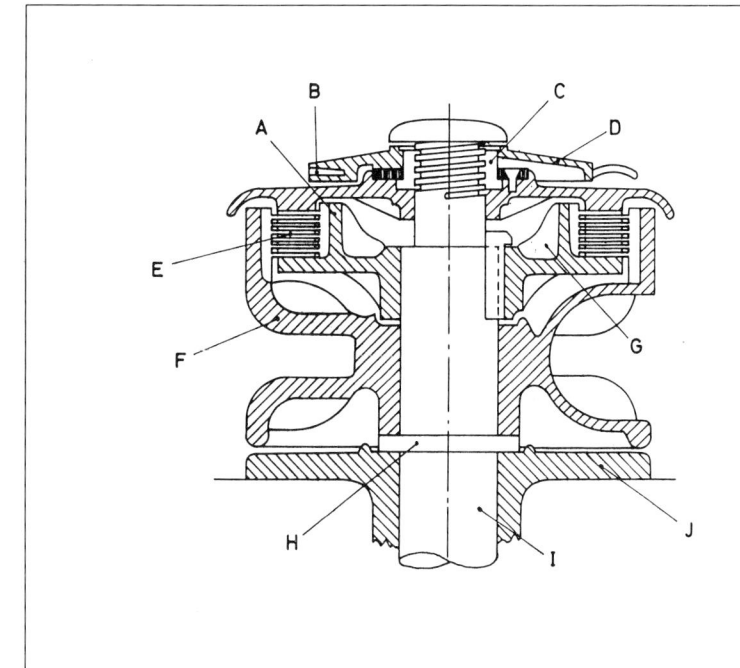

Figure 8. Harfield's original pattern cable holder
A. Casting keyed to spindle
B. Slot for portable operating lever
C. Nut
D. Crown plate
E. Friction plates
F. Holder
G. Six webs on casting 'A' holding one set of friction plates
H. Collar taking the weight of the holder
I. Spindle
J. Bed

Figure 9. Bonnet
A. Sliding shutter in raised position
B. Sliding shutter in secured position
C. Cable link in secured position
D. Slide
E. Deck pipe

to the driving spindle by means of a multi-plate friction clutch, which could also be used as a brake when veering.

Capstans and cable holders were now on the open forecastle and consequently the deck pipes, which were now beginning to be called navel (not naval) pipes, had to have the ability to be closed to prevent the ingress of water. This was done by means of a portable 'bonnet' (Figure 9) which could be closed by means of a sliding shutter, shaped to fit around the cable.

Working cable – anchoring

Cable work with close stowing anchors was basically the same as that for stocked anchors with chain cable. On approaching an anchorage the anchors were cleared away for letting go, leaving them suspended by the tumbler chains. Any lashings on the cable were taken off, the chain lockers cleared away, the compressors 'put back' or 'backed' (released) and the bucklers raised. If the chain lockers were well abaft the bitts a small amount of cable was ranged, sufficiently heavy to draw the cable from the chain locker as it ran out through the hawsepipe.

In destroyers and other small vessels not fitted with cable holders the cable was then bitted to control it as it was veered (Figure 10). In ships with cable holders the cable was veered under their control. (When ships were first fitted with cable holders it remained general practice to bitt the cables for veering, but when the later ones with more

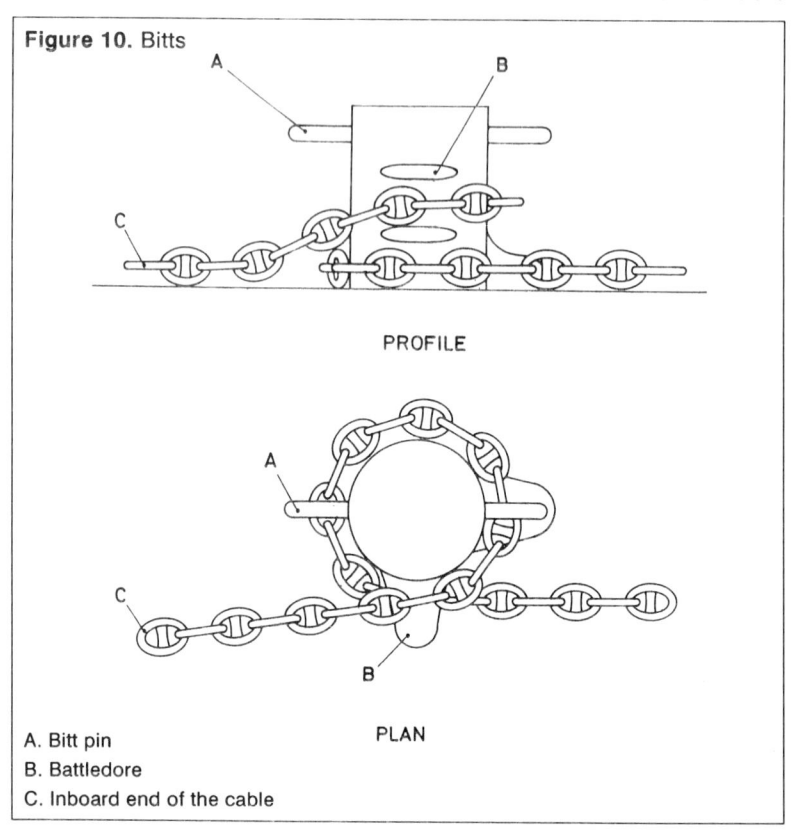

Figure 10. Bitts
A. Bitt pin
B. Battledore
C. Inboard end of the cable

PROFILE

PLAN

42 MODEL SHIPWRIGHT 93

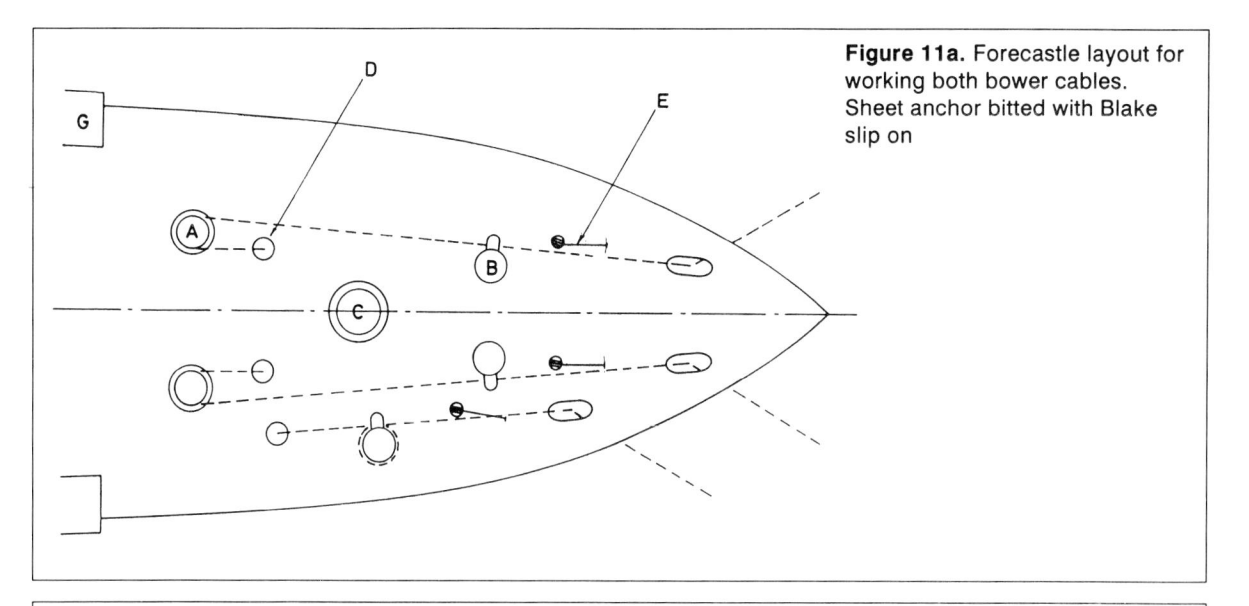

Figure 11a. Forecastle layout for working both bower cables. Sheet anchor bitted with Blake slip on

11b. Forecastle layout for working the port cable with the capstan and the sheet anchor with the starboard cable holder.

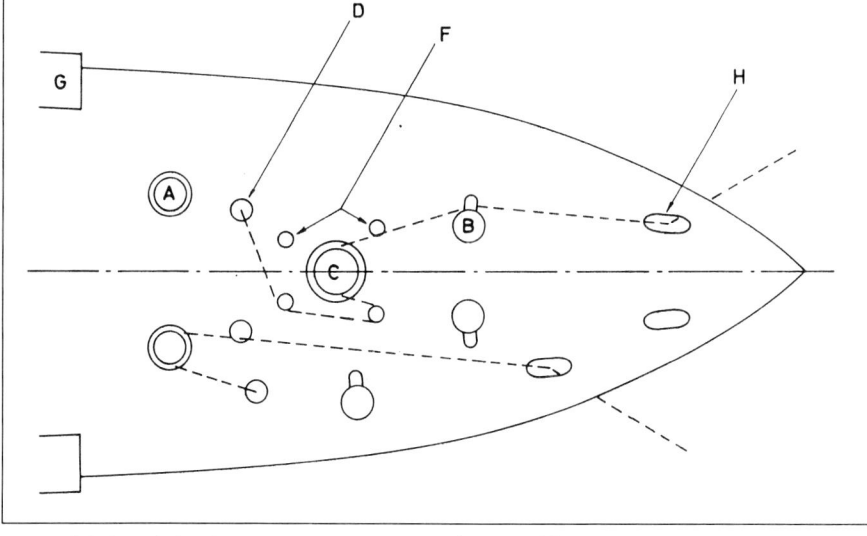

A. Cable holder
B. Bitts
C. Capstan
D. Deck pipe with compressor below
E. Blake slip
F. Portable rollers
G. Anchor bed
H. Hawsepipe

powerful band brakes came into service the use of the bitts for braking fell into disuse.)

After the anchor had been let go and the ship had come to it, the slip stopper was put on as a preventer. Cable was then veered until the slip was just taut and the compressor was bowsed tight. Finally, the cable holder brake was screwed on tight and disconnected. The ship rode to this, or the bitts if the cable had been bitted.

Working cable – mooring

Mooring ship in this era was exactly the same in principle as when stocked anchors were used with chain cable. (See Part II of this series.) It was only the hardware that differed.

Working cable – weighing

Cable was normally weighed using the cable holders but the capstan could be used if required. As it was ahead of the deck pipes portable vertical rollers were placed in deck sockets on either side of it to lead the cable almost half a turn around the barrel to give a good grip (Figure 11b).

The method used to cat and stow a vertically-stowed anchor was similar to that used for a stocked anchor. In place of the old wooden cathead ships now had a 'clump cathead', which consisted of a large metal sheave set at an angle in the deck edge (Figure 4). When the anchor had been hove up so that its ring was just below the hawsepipe the cat pendant was taken forward from the capstan over the sheave of the clump cathead and hooked or shackled to the ring of the anchor. Then by veering the cable and heaving in on the cat pendant the anchor was placed on its chock. Here it was secured by two tumbler chains in much the same way that anchors were secured on a horizontal bed, except that in this instance the tumbler bolt was vertical, not horizontal and was released by a lever on the upper deck. Extra securing chains were passed, including one through the ring, fitted with a screw and hand wheel tensioner. When all was secure the cable was hove up tight against the thumb cleats on the ship's side, where these were fitted, lashed down wherever possible, the bonnets secured and caulked to keep out the sea.

With ships fitted with horizontal anchor beds the procedure was somewhat different. In place of the old cathead, ships now had a 'cat davit' (Figure 2) which could be hinged down flat on the deck when not in use. It was fitted with two metal blocks, one at the head and the other shackled to an eye on a band which was free to revolve around the davit just above the hinge. Prior to weighing a wire cat pendant was rove through these blocks and up through the hawsepipe from outboard in and its end secured temporarily.

Each bower anchor was fitted with a gravity band to which was shackled a 'ground chain' long enough to reach to the deck inboard of the hawsepipe with the anchor in sight. This was stopped to the cable along its length and its inboard end was secured to the cable with a small slip. With the anchor high enough for the end of the ground chain to be inboard the latter was unslipped and shackled to the cat pendant. The other end of the cat pendant was then taken to the capstan and the pendant hove out through the hawsepipe, breaking the stops as it did so. Once it was clear the cable was veered and the cat pendant hove in until the anchor was at the davit head. The davit was then eased forward using its guys and with the help of a couple of jiggers the anchor was lowered onto its bed and secured. As with the vertically-stowed anchor, all that then remained was for the cable to be hove up tight against the thumb cleats (Figure 12) and everything secured for sea.

The ground chain was very much the invention of the 'timed evolution' Victorian Navy. Its sole purpose seems to have been to obviate the need for men to go over the side of the ship to secure the cat pendant to the gravity band and thus save time. However, as the cable dragged over the seabed the stops were frequently chafed through and the slip disengaged. (This has been recorded as happening as frequently as on two out of three occasions.) Dealing with an anchor in that state took a considerable time and defeated the whole reason for the existence of the ground chain. Consequently, it was ultimately abandoned and the anchor catted in the same way as the vertically-stowed one, but using the gravity band not the ring.

to be concluded

Figure 12. Outboard lead of cables as secured for sea showing three alternative layouts. The upper sketch shows the most common method. Both this and the middle sketch indicate that the sheet anchor is stowed on the after bed. The lower sketch shows a simpler lead for the cables but the sheet anchor is now on the forward bed.
A. After anchor bed
B. Cat davit
C. Forward anchor bed
D. Thumb cleat

Coringa on trials in the Firth of Clyde.
Photograph: Courtesy of Dumbarton District Libraries

MODELLER'S Draught

STEAM TUG *CORINGA*

Specification

Length: 134ft 4in
Breadth: 25ft 1in
Depth of hold: 13ft 0in
Draught: 12ft 2in
Tonnage, gross: 287 tons
Displacement: 665 tons
Machinery: Triple expansion, 4-blade propeller
Speed: 13 knots

The tug shown in the plan is slightly unusual in that she has twin funnels, not all that common although a number of early vessels were so fitted, as were some of those owned, among others, by the Dutch firm Smit, and, of course, there were the Admiralty *Rollicker* class, and the well-known *Foundation Franklin*.

Coringa was built in 1913 by Wm Denny Bros, Dumbarton, for the British India & Queensland Agency Co Ltd, London, for operation as a salvage tug based in Brisbane, Australia, to which country she sailed under her own steam.

Modelling notes

The plan as reproduced is for a model to $^3/_{16}$in = 1ft scale (1/64) which would have a displacement of about 5½lb (2.5kg) to the load waterline. If enlarged to 1/48 scale, giving a model about 36in long the displacement would be about 13lb (5.8kg), a useful yet manageable weight, and the model of a size that would not present any undue problems for transport by car. Reasonable access to the interior of the hull could be gained by making the casing and skylight removable in one piece.

There is nothing untoward in the tug's hull. The lines show it to have fine lines with low freeboard. The keel rake is not excessive, but the open type stern frame should be noted. Bilge keels were fitted as shown, and there is a heavy timber fender all round at deck level, as can be seen on the cross section detail, which also shows the strakes of shell plating.

On the deck plan the engine casing is shown extending forward, with a narrow cross passageway between it and the forward deckhouse. There are two coal hatches and a central door to a store at its fore end. The forward deckhouse has a central door in the after end. However, the bridge deck over is continuous with access to it by way of ladders to the bridge deck wings. The towhook is fitted at the after end of the casing, with a vertical supporting pillar beneath it. The steering engine is fitted in a compartment at the after end of the engine-room skylight, with the steering chains passing from this compartment out to the inside of the bulwarks, and thence aft to the rudder quadrant; buffer springs are fitted to the chains as shown. A grating is fitted over the top of the quadrant, supported by an angle iron framework. There is a steam-driven capstan aft, and salvage pump connections are on the top of the low casing on the after deck. The forward section of the upper deck is planked, the rest is bare steel. Two tow beams are fitted across the after deck, bracketed to the top of the bulwark.

Two lifeboats are carried under radial davits, and stowed on a wood sparred deck extension P & S to the casing. A 16ft dinghy is stowed on chocks on the starboard side at the after end of the casing. A permanent wooden awning is fixed over, and a few inches above, the top of the wheelhouse. The steering wheel is in the open, with a full height screen with windows at the fore end of the deck, with provision for canvas dodgers to be fitted on either side of the screen out to the bridge wings. The bulwark round the bridge deck is planked vertically. A galley funnel runs from the after end of the house on the bridge deck to the forward funnel.

Colour scheme

Hull: black, narrow white ribband at deck level; red below waterline.

Casings, engine-room skylight, inside of bulwarks, coal hatches, skylights: teak colour.

Curtain plate of navigating bridge and of boat deck: white. Supporting stanchions to these decks white.

Deckhouse on bridge deck, screen forward, bridge deck bulwark: varnished teak.

Lifeboats, davits: black, boat covers grey canvas. Sparred deck under boats light grey.

Windlass, capstan, anchors, anchor davit, quadrant, steering chains, salvage pump outlets, bollards, fairleads, timber heads, towhook, tow beams: black.

Bare steel deck: red oxide.

Ventilators: black, inside cowls red.

Mast: varnished wood.

Grating over quadrant: varnished wood.

Funnel: black, two white bands divided by narrow black band.

THE WOODEN WATERLINE

by H G Harris

'Why don't you make a ship model for him?' This suggestion was made by my wife one evening when we were discussing an impending visit to Canada, for there was no doubt that forthcoming hospitality required some reciprocal gesture to our host. The idea fitted well with his interests, but a ship model as normally understood was a pretty tall order for me in anything less than a couple of years. However, something had to be done.

The question of what ship to make more or less solved itself, for I had a copy of the splendid set of drawings of the schooner *Bluenose II*, prepared by Commander L B Jensen (see *MS* 21). Being a famous Canadian vessel, it could hardly fail to please a Canadian citizen.

The form of the model was another matter. As time was short a fully detailed model was out of the question. In the end a small half model seemed to be the answer, on the grounds that a half-hull would be quicker than a full one, and by tradition could be simple and devoid of fiddly detail. At a scale of 1/192 the hull would be about 9in long, so transport would be easy. Nevertheless, a half model by itself seemed a bit stark, and modelled rigging would take far too long. So why not mount the hull on a drawing of the rig and put the whole thing in a picture frame? No sooner thought than decided.

Not having carved a hull from the solid before, a trial run was done on a piece of yellow pine. As the final shape started to emerge it began to dawn on me that eventually a horizontal white line would have to be painted around some quite complicated shapes at bow and stern, and the likelihood of achieving a crisp straight line with masking tape seemed very remote. The problem worried me very much, and the answer came to me in a dream – such are the compulsions of model shipbuilding.

It was, of course, perfectly simple – make the hull of a sandwich of three pieces of wood, the uppermost being the black topside, the middle the white waterline, and the bottom piece the red underwater body. I made up a rectangular block using sycamore for the top and bottom pieces, with a piece of 2mm three-ply forming the meat in the sandwich. This latter seemed poetically appropriate for what was to be, in effect, a three-ply ship, and convinced me, quite irrationally, that I was on the right lines.

The relationship of the three layers was preserved during and after carving by two vertical dowels. A central screw held all together and allowed for the block to be taken apart prior to painting. To hold the assembled block while carving, the flat back was secured to a baseboard by horizontal screws engaging with the top and bottom pieces of sycamore. Carving then proceeded as before, but using the wooden waterline as the vertical datum when applying the section templates. One of the photographs shows the three pieces separated at the completion of the carving stage.

The hull was reassembled and sprayed with car body primer filler, which was rubbed down between coats with very fine 'wet and dry' abrasive paper used dry. The finished surface was excellent, but would the filler crack off at the edges when the hull parts were separated? Happily, no such thing occurred, and each part was painted in its appropriate colour, and the model reassembled when dry. The result was amazingly good, and far better than I could have achieved by other means.

I had the drawing of the fully-rigged ship photocopied at the appropriate scale on tinted paper, and coloured in the masts, spars, flags and certain other features with water colour paints. The print was mounted on to hardboard and the hull screwed on in the right place. The model had to be located very accurately on the drawing, but as my wood carving precision is not of the highest order I took the precaution of blanking out most of the hull outline when making the photocopy, which prevented odd bits of the drawing appearing beneath the hull. A picture framer designed the mount and framed the whole assembly; no glass was fitted.

The end result was very satisfying and warmly received by the new owner. I am convinced that the extra time taken to make up the three part block was more than recovered by the ease and certainty with which a good multi-colour paint finish was obtained. Although this method is original so far as I am concerned, I cannot suppose that others have not adopted so simple a solution to this problem, but I have not seen it described before.

I am indebted to Mr L B Jensen for permission to reproduce the rigging plan of *Bluenose II*. ❑

[The set of drawings mentioned above have now been reproduced in the book *Bluenose II – Saga of the Great Fishing Schooners*, by L B Jensen, published in 1994 by Nimbus Publishing Ltd, PO Box 9301, Stn A, Halifax, Nova Scotia B3K 5N5, Canada. ISBN 1-55109-063-5. Copies can be obtained through booksellers in this country. Editor]

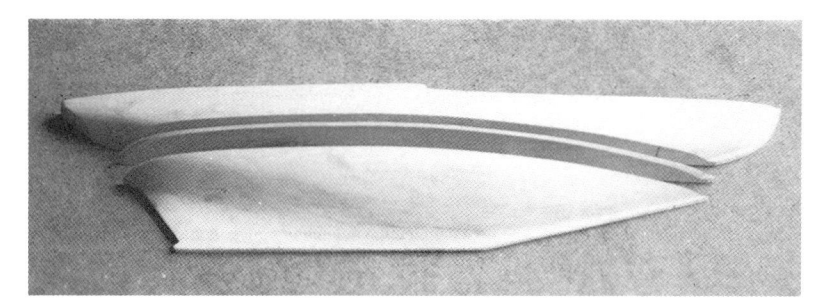

The hull separated into its three parts after carving.

The finished hull mounted on the rigging plan.

The model mounted in the picture frame.
Photographs by the author

The 1/120 model of the *Stirling Castle*.

Stirling Castle (1936)

The Union Castle liner

by Ronnie Lawson

I have always been interested in ships and ship models, but it was not until I retired some years ago that I was able to find time to devote to this interesting and absorbing hobby. I set about obtaining as much experience as I could by building several models of ships to my own designs, some of which were radio controlled. Not only did this give me valuable experience in building methods but led to a study of as many books and as much information concerning ships and their design and construction as I could lay my hands on. This was essential for, apart from the pleasure gained from building the models, I get a great deal of satisfaction in drawing their plans, and to be successful these had to be based upon first class research.

Stirling Castle was my first attempt at a scale model of a prototype vessel, something which I would have been unable to attempt without an understanding of what was involved, particularly in relation to the complexity of this large passenger liner. The choice of subject was based on nostalgic memories of seeing her in the late 1930s together with the fact that in my view she was a very handsome and powerful looking vessel.

It took about twelve months of research before I had obtained all the information I required to start preparing my own set of working drawings. I had been very fortunate in being able to obtain a rigging plan and the general arrangement plans of the upper decks of the ship from her builders, Messrs Harland & Wolff Ltd, Belfast [*Note: All Harland & Wolff's plans are now held by the Ulster Folk Museum, but unfortunately it is not possible to obtain copies of any of these plans. Editor*].

Why did I undertake the lengthy task of drawing up a full set of working drawings? Because I have found that shipyard plans have a number of drawbacks. Rarely do they show end elevations of deckhouses; often they do not include on the profile an elevation of many of the deck fittings shown in plan view on the deck arrangements, and a lines plan, as was the case with the *Stirling Castle*, is more or less unobtainable. So I produce my own plans at a more convenient scale [*a shipyard plan at 1/48 scale, some 12ft*

The hull under construction, with its assembled keel, frames, and stringers almost ready to have the shell plating fitted.

Preparing the margins and deck planking of E deck on the drawing board.

to 14ft long, perhaps more, is not unknown. Editor] and, by dint of much research, include thereon all the information I require. Originally, I had intended to do this set at 1/144 scale, but the purchase of a hatchback car made me realise that the transport of a 6ft model would be possible, and so I decided to build to my usual scale of $^1/_{10}$in = 1ft (1/120).

Stirling Castle was built in 1936 by Messrs Harland & Wolff Limited, Belfast, for the Union Castle Steamship Company's mail service between Southampton and South Africa, and left on her maiden voyage on 7 February 1936. She survived the War and was sold to Japanese shipbreakers in 1966. Her sister ship *Athlone Castle* was completed a few months later and she, too, survived the War and was sold to Japanese shipbreakers in 1965.

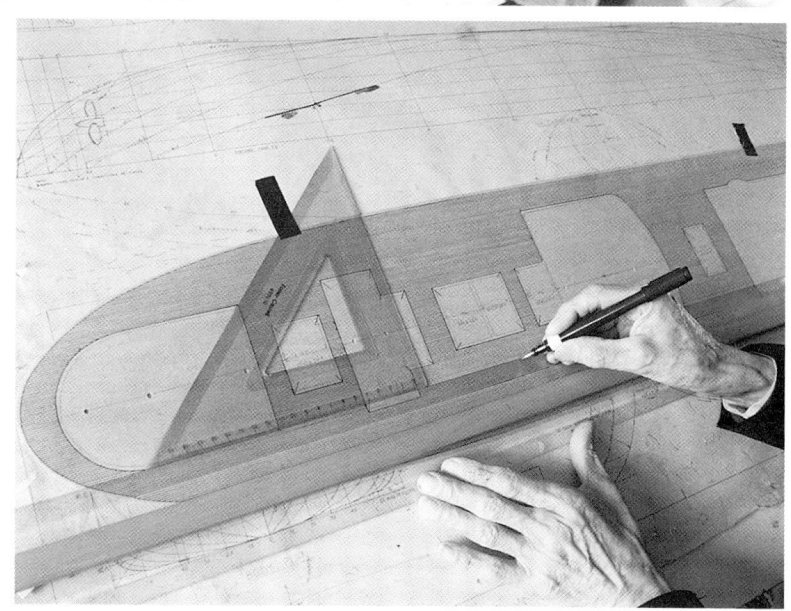

Specification

Length overall: 725ft 0in (220.98m)
Length BP: 680ft 0in (207.26m)
Breadth mld: 82ft 0in (29.99m)
Draught: 31ft 10in (9.70m)
Gross tonnage: 25,550 tons
Machinery: Two 10-cyl H & W Burmeister & Wain oil engines; 24,000bhp.
Speed: 20 knots
Passengers: (as built) 1st 291, Cabin 492

Bridge front. The curved parts of the front (the 'panels' in the text) had to be steamed to shape after the window openings had been cut out.

The model

At the scale of 1/10in = 1ft the model was 72½in long overall by 8.2in beam by 3.18in draught. Originally, I had intended that she would be a radio-controlled working model, but as work progressed I realised that she would be far too heavy for me to deal with at the pond side, particularly when fitted with ballast, batteries and R/C gear. So I decided to make her just as an exhibition model.

When starting on the plans, I began by preparing a lines/body plan. I had several photographs of the ship, and a photocopy (obtained from the Science Museum, London) of the special edition of the journal *The Motor Ship* published in February 1936. This contained a description of the ship, plans of some of the accommodation, and a number of photographs including some taken while the ship was under construction on the building berth. Unfortunately, I was unable to obtain any general arrangement plans of the lower decks. If I had, my task would have been much easier. The rigging plan gave me the contour of the bow and stern and the sheer of the uppermost continuous deck, D deck, while the deck arrangements gave its outline. The only other accurate drawing I had was a cross-section in way of the engine room, which gave the rise of floor. However, by using such information as I had, coupled with a careful study of the photographs of the ship in frame on the stocks, and some aerial shots of the ship under way, which gave me an idea of the length, shape and form of the entrance and run, I was able, after much trial and error, to draw up a body plan and a lines plan. I knew that such limited information could not possibly produce a strictly accurate lines plan, but there seemed to be nothing for it but to press on, and I feel that, with all due modesty, the result has produced a very reasonable looking hull form.

I had decided to build the hull using plank-on-frame construction because of its length. My previous models had all been built on the bread-and-butter principle. Being drawn full-size for the model, I was able to trace the outline of the body plan sections on to 1/4in thick marine plywood to make the frames. The keel was an 'I' section girder made up from 1/4in marine ply, while the cruiser stern was carved from jelutong in two halves. The frames were slotted into the keel. The shell was made from 1/16in thick ply cut into long strips to follow the sheer, pinned and glued to the frames with Cascamite. A few stealers were required at the after end. The interior up to load waterline level was treated with glass-fibre, for at this stage I was still looking upon her as a working model. The hull was now plated with 1/64in ply arranged from photographs to represent the run of the outer strakes of shell plating.

Upper promenade deck showing bay windows. The windows of the lower deckhouse had curtains painted on the inside of the glass.

Before drilling the hull for sidelights it was first sprayed with primer filler, a cellulose based heavy primer only possible to use with a fairly powerful spray gun. This was rubbed down with 600 grade wet and dry paper used wet to give an excellent finish to the base coat. The hull colours, Humbrol paints mixed to the appropriate shades, were brushed on. When all the coats had been applied the holes for the sidelights were drilled and reamed out to size by hand. Strips of transparent acrylic sheet, cut into lengths, each long enough to cover a number of sidelights, were fitted on the inside of the hull, being glued between the holes. *Scale-Link* photo-etched letters were used for the ship's name and port of registry. The letters were superglued in place, painted gold, and shading added under each one with a draughtsman's fine pen to give the effect of a black shadow. The draught marks were minute slide-off number transfers obtained from a model railway shop. The narrow teak moulding dividing the lavender colour of the hull from the white topside was made from 00 gauge model railway line ('I' section). A considerable number of $1/32$in brass wire pins were soldered to the inside of the rail which was then pressed into holes drilled in the hull. The pins were completely invisible when filed flush with the outer surface of the rail, and so no adhesive was needed to fix the moulding to the hull; it was painted before being fixed in place. When everything had been finished the hull was given a coat of Satincote varnish.

Decks

All the exposed decks were planked, except for a few small areas of bare steel. Although the width of the planks on the ship was 5in, I settled for 6in or 1/20in scale size because this was easier to measure. Each deck was made of selected and well prepared ply, cut to size and shape according to its position. The decks were taped to the drawing board and the margins, including the waterways, lightly marked in with pencil. The longitudinal centreline of each deck was established, and the lines of planking drawn in with black ink using a very fine draughting pen (so as not to over emphasise the caulking) and the parallel motion straight edge fitted to the drawing board. There was no indication on the plans of plank lengths, so I marked the butts according to the usual shift of butts layout, ie every fifth plank butt in line, making indi-

Details at the after end of the boat deck and on the deckhouse forming the Cabin Class lounge. Note the stack of liferafts stowed on deck below the lifeboat.

vidual plank lengths about 25ft. The line of the waterway was marked in at this stage, and where the plank ends were joggled into it this was simulated by gently scraping away the already marked plank ends on the waterway fore and aft with a razor-sharp blade, taking care not to spoil the surface of the ply deck.

Deckhouses

The various deckhouses were made of $1/32$in ply, steamed and curved to shape where appropriate after the openings for the windows (about 384 of them in all) had been cut. The surface of the ply was treated with sanding sealer and sanded down several times after which the panels were painted. The house sides and ends had to be cut to conform to the sheer and camber of the decks. The $1/10$in deep skirting along the base of each house was marked and filled in using a thick pen point in preference to a brush since this was easier to control. Window frames, being slightly larger than the cut openings, were drawn in and then filled in, along with the bare wood cut edge of each opening. To facilitate their removal when necessary,

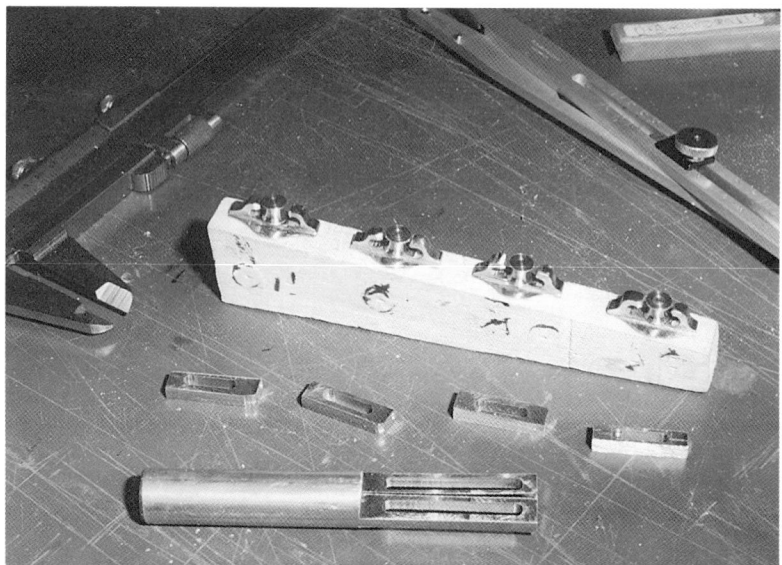

Single roller fairleads under construction; some were made from the brass bar in the foreground.

the locations of the houses on the decks were marked with either $1/32$in brass wire pins or pieces of $1/8$in wood dowel.

Most of the doors were cut from

Finished single and triple roller fairleads ready for painting.

transparent acrylic sheet painted beforehand on the underside, since no matter how roughly (within reason) paint is applied to the underside, when viewed from the outside it shows a perfectly covered and smooth surface without brush marks. The small hole for the door handle was drilled in each one before it was fitted to its deckhouse. Other details, such as storm rails, were added as work progressed.

The most difficult house to make was than on E deck, which had a number of bay windows. The windows in this house were different from those elsewhere and were modelled from micro-strips with individual glazed panels. At the forward end of this house were a number of large arched windows in the area of the First Class lounge, which were similarly modelled.

The curtain plate along the edge of these overhanging decks was made from $1/16$in thick ply and glued in place. $1/16$in ply was used so that holes could be drilled along the top edge to take the guardrail stanchions, of which there were a considerable number. On the promenade decks the stanchions had to be fitted *in situ* and spaced in conjunction with the deck support stanchions, some of which were aligned with davit trackways on the boat deck. Many of the guardrails were of the four-bar plus teak-top-rail type, and in places the deck support stanchions appeared to pass through this teak top-rail [in practice it was cut and fitted round the stanchion. Ed]. The teak top rail was made of brass, notched in way of the deck support

Plug, mould and shells for 30ft lifeboats. Surplus plastic will be cut off with scissors before shaping the gunwales. Note the screw in the plug to act as a handle.

stanchions, and with holes drilled in it through which the $1/32$in wire deck support stanchions were pushed and soldered. The deck support stanchions under the davit trackways were cut from 'I' section 00 gauge railway line. The bars for the rails were made from 0.3mm straight brass wire, soldered to the rail stanchions, all of which had been fitted into holes drilled in the top edge of the curtain plate and soldered to a hole drilled in the top rail. Once the flux had been cleaned off the soldered joints, the inside of the rails was painted white. Hence the reason for making the upper decks detachable in the early stages.

Deck fittings

I find that some of the most interesting and enjoyable items to make are the deck fittings such as bollards, fairheads, derrick crutches, winches, ventilators and casings, and so on. Although a lathe is not strictly essential, and much good work is done using nothing more than an electric drill, having one does make the job easier. Rather than describe how all the fittings were made, I will mention some of those where a lathe and its attachments was very useful.

The ship was fitted with single, double and triple roller fairleads, as well as some plain ones. The most difficult parts to make were the horns at each end of the fairlead. I tried, without success, to make them from plastic card, and by making patterns and moulding them. In the end I made them from solid brass bar, machined flat to the scale thickness and then milled with a slot the length of which was equal to the distance between the horns. After cutting off the top of the slot the outer ends of the horns were machined to a larger radius, the top corners filed to the correct shape. These were soldered to a baseplate cut to shape from 0.015in brass plate, with holes drilled to take the pins of the rollers.

The rollers were turned on the lathe from brass rod using a form tool ground to the shape of the waist of the roller. A 0.8mm diameter hole was drilled through the centre of each roller to take a $1/32$in diameter pin (which was a tight fit in the hole). By this method each of these items was identical in size, shape and radius. They were painted matt black.

Specially ground form tools were also used to turn up the winch drums and barrels. After each had been drilled to take the $1/32$in diameter shaft, the outer end of each drum was machined out to form a hollow recess. The winches were painted black, with the recess in the drum

Deck forward showing rigging, shrouds, ratlines and variety of deck fittings.

The after deck.

Lifeboat shells fitted with rudder and propeller.

ends white. As there is much repetitive work on a vessel of this type, and to a lesser degree on other types, it is important to keep a log of the various micrometer settings, etc on the lathe, to ensure uniformity when machining the different items.

The cowl ventilators had mainly circular tubes which were made either from brass tubing of scale size, or from brass bar drilled out and turned to size. The cowls were of brass, soldered to the tubes. Form tools were made to produce the cowls, the insides of which were machined to produce a thin shell. On some ventilators the tube was only 0.075in diameter, with cowls 0.15in diameter.

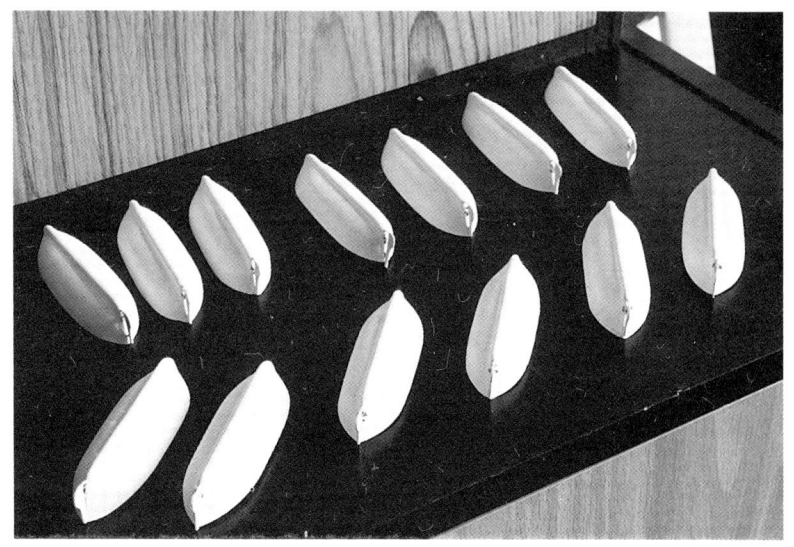

Some years ago I made a worm-driven dividing head for making small gears and circular items which had to have perfect, evenly-spaced holes or teeth. This device was invaluable when it came to making the gears for the windlass. The largest gear wheel, as well as enabling the teeth to be cut round the rim, had to have eight evenly spaced holes cut through its face. Another example where accurate dividing was essential was with the two operating handwheels on the windlass braking mechanism, each of which was fitted to a vertical spindle just behind the windlass frame. I made these from $3/16$in brass bar. After drilling a 0.8mm-diameter hole down the centre, each wheel was parted off at the correct thickness, and mounted in turn on the dividing head to have six evenly-spaced holes drilled around the centre and through the face. I doubt if I could have made the two ship's wheels without this head. They were made from 0.375in diameter brass rod with eight holes for the spokes drilled through what was to become the rim. After machining out the middle they were parted off to the correct width. The boss would have to be no more than 0.125in diameter, which presented difficulties when drilling out eight holes to take the inner end of the spokes, which kept running into one another. In the end I just made the spokes exactly the same length from rim to boss, their ends thereby producing a 'hole' in the centre since they lay so close together. The boss was very carefully centred on a jig and then glued into the hole with superglue, taking care to make it appear concentric to the rim.

The anchor cable was cut to the required length from what originally was a very cheap chain intended presumably to be worn on the wrist. This was stretched out taut on a piece of scrap board and held by a pin in each end link. A stud for each link was cut from 0.013in brass wire and fixed across the centre of each link with superglue. The finished chain was painted matt black.

On a photograph of the ship in the *Motor Ship* article the hatch tarpaulins appear to be white. However, on the model I fitted the hatches without covers to show the individual hatch boards.

Modelling the funnel

The proportions of *Stirling Castle*'s funnel seemed to add a sensation of great power to an already fine looking vessel. I began by cutting a plastic card template to the shape of the funnel in plan view, and used this to mark out the shape on two pieces of ½in thick balsa wood, to form the top (the plated over top area of the funnel) and bottom of the funnel. These were joined by a number of pieces of ½in thick balsa cut to the correct rake, to locate the top piece in its right position, some 4ft 6in below the top edge of the funnel. This was not shown on the plan, but a photograph of the funnel waiting to be lifted on to the ship showed two men standing in the funnel and peering over the rim, from which I estimated the depth of the plated top below the rim. The frame was covered with $1/64$in ply steamed and fitted round the framework and glued in place, with the butt coming on the after end. Any slight gaps in this were filled and the whole sanded smooth. The rim round the top edge, and the lower bands were narrow strips of thin plasticard glued just above some masking tape (used to help position them). After gluing two doors 6ft high on the fore end of the funnel, it was spray painted in the company colours. This was followed by fitting the whistle, two sirens and their associated ladder and platform. A number of chain links hung down at intervals round the top of the funnel below the rim band. Six funnel guys, three each side, were attached to eyeplates on the lower band. These wires were secured to eyeplates on deck with rigging screws. Various exhaust outlets were fitted to the plated top of the funnel.

Lifeboats

The ship carried sixteen lifeboats, two of which were smaller emergency boats and one a transom-stern motor boat. These were mass produced. A pattern was carved in jelutong which, after being smeared with Vaseline to act as a release agent, was used to make a mould out of plastic padding. The pattern was then reduced slightly in size to accommodate the plastic card which

The stern, showing propellers, shaft bossing and rudder.

would be heated to form the shell of the boat. A piece of 0.020in plastic card was fastened with drawing pins to the rectangular wooden box holding the mould, then heated by being placed under the cooker grill. When ready it was removed and the pattern (plug) pressed into the mould. Surplus material was trimmed off with scissors. Each boat was fitted with a rudder, small three-bladed propeller, and canvas boat cover. The propellers were made of brass bar machined to about $1/10$in diameter and fitted to a short length of $1/32$in diameter rod to represent the propeller shaft. This was held in a pin chuck while the three blades were ground out and shaped using a Minidrill and cut-off disc.

The davit trackways were made from pieces of extruded 'U' section brass strip soldered together and to their supports on a jig to ensure uniformity, and fitted and secured in place on the model with a large triangular-shaped bracket plate at the lower outboard end. The boat cradles were cut from plastic card and fully detailed. No sheaves for the lowering gear were visible, these being within the cradle frame. The cradles were secured to their trackways and the boats fitted.

Midships, looking forward.
Photographs by the author

Masts and derricks

The masts, derrick posts and derricks were all made from brass, as were their various fittings. Having a copy of the builder's rigging plan was of great help when it came to rigging the model. Perhaps the worst aspect of this was the ratlines. Initial efforts with nylon fishing line and polyester sewing thread were disastrous. Eventually I used 0.3mm brass wire in straight lengths, soldering the ratlines to the shrouds on a jig. When finished they were painted black. On the ship the shrouds were of galvanised steel wire rope, and the ratlines were galvanised steel rods.

To build a large model of *Stirling Castle* presented a challenge, and I am glad that I accepted it; certainly it will not be the last vessel of this type that I shall build.

Colour scheme

Hull: officially known as French Grey; lightish grey with a touch of lavender. White above teak moulding on top of rail and top of bulwarks on D deck. Below waterline red (Humbrol 19) with a touch of black and satin varnished).

Superstructure and deckhouses: white, but specification indicates ivory on promenade deck houses only.

Masts and heavy derrick: reddish brown.

Derrick posts, derricks: white.

Hatches: black on model.

Lifeboats: white, canvas covers Humbrol 145 medium grey.

Ventilators: white, inside cowls red.

Bollards, fairleads: black.

Windlass, winches: black, inside of winch and windlass drums white.

Funnel: orange-red with black top.

Decks: planked, bare steel decks black.

Swimming pool: no information about colours, but painted grey on model.

References

Copies of shipbuilder's plans.

Article in *The Motor Ship*, Volume XVI, No 193, February 1936.

Official photographs of the ship, from Ulster Folk and Transport Museum, Holywood, Northern Ireland.

Acknowledgments

I am indebted to Messrs Harland & Wolff's Technical Services Department, Belfast, for much help with this project, also to Mr Jimmy Wood for advice on how to build a plank-on-frame hull, to Mr W A Richmond, deck officer on the ship in 1939, for advice and loan of personal photographs, and to Mr Colin Bishop for loan of photographs. ❏

The model was awarded a Silver Medal at the 1995 Model Engineer & Engineering Exhibition in London. *Editor.*

'Vengeance'

A whale of a diorama

by Scottie Dayton

The diorama of the sperm whale tossing the 30ft whaleboat into the air. Note the shark's fin just off centre at the bottom.

Larry Hubbell, from Winnecta, Illinois, was to find that his artistic background in commercial art stood him in good stead when he decided to try his hand at model shipbuilding in 1971. Before long those artistic talents led from the pure reproduction of scale ships to presenting them in highly original situations.

His first essay in this direction occurred in 1975 when he realised that 'most people do not know what the inside of a ship looks like, and neither did I at that time'. Following extensive research into eighteenth-century English men-of-war he produced a cutaway of a First Rate in which every nook and cranny was exposed, and complete with all the signs of habitation. This remarkable model, valued at over $60,000, now resides in the Navy War Memorial Museum, Washington, DC. Its success launched him in a new direction. In future his models would depict a particular moment in a ship's history.

His next masterpiece was a 1/32 scale diorama of Lord Nelson falling at the Battle of Trafalgar. One hundred and fifty-four converted military figures were used to reflect the activities on board HMS *Victory* at the moment the French musket ball struck Nelson. This diorama, housed in a case 27½in high x 21in wide x 30in long, is highlighted by small spotlights hidden in the sails and by lanterns hung along the gun decks. It, too, is now in the Navy War Memorial Museum.

This was followed in quick succession by a 1/75 scale diorama of the English First Rate of 1719, HMS *Royal William*, and her crew sailing through a choppy sea, and one at 1/24 scale of the sinking of the Swedish warship *Wasa*. The project for 1990 would be the creation of a whaling diorama to answer the ques-

Cutaway model of an eighteenth-century English First Rate. Woods used in its construction included apple, boxwood, cherry, maple, poplar and walnut. Scale 1/64, length overall 42in.

The scene on board HMS *Victory* at the moment Lord Nelson was struck by the musket shot.

tion 'what happened out there?'.

He began the necessary research by reading numerous books and studying paintings of all styles and descriptions relating to the subject. The first problems he encountered lay with the inconsistencies of dress, even down to braces, hair styles and beards. Moreover, he had never seen a diorama of a sperm whale upending a 30ft long whaleboat and scattering its crew. Certainly whales had upset boats and tossed men into the sea, but what did a sperm whale look like when so doing?

For the diorama he decided to

MODEL SHIPWRIGHT 93 59

Close-up of the scene on *Victory*'s quarterdeck in the other photograph after Lord Nelson had been hit.

work to a scale of 1/32, so that he could use Airfix multi-pose military figures of the same scale. These are his favourite plastic men, because their interchangeable arms, legs, torsos, heads, etc make them easy to pose. To facilitate construction he purchased a whaleboat kit for its plans, frames and planks from the Laughing Whale Company. The hull was an inch too short for a scale 30ft boat, so he compensated by running the lapstrake planks ½in beyond the bow and stern frames, then gluing and clamping the ends together.

Next day one of his modelling friends dropped by and asked, 'isn't there any sheer to a whaleboat?' another friend coming for a visit observed, 'your boat looks too skinny – like a canoe!' That did it. Deciding that a sperm whale could cause more damage than originally planned, and with his second friend as a spectator, he placed his thumbs on either side of the hull at amidships and pulled. Amazingly, the boat did not snap or crack, attesting to the strength of the applewood and Elmer's White (PVA) glue. Instead, a new sheer and beam were formed which matched the plans. Quickly inserting a thwart to hold the hull's new shape, now perfect in every way, he solemnly announced to his friend that 'he had planned it that way all along!'.

In the diorama, the impact of the whale smacking the bottom of the hull with its head has sent splintered boards flying and left a huge hole in the bow. A metal patch on the bottom of the hull is evidence of a previous mishap.

To create the sperm whale he began by rolling up newspapers to its basic form and wrapping the bundle with masking tape. Celluclay (commercial papier-mache sold at most hobby stores) was mixed with a generous amount of Elmer's glue and applied no more than 1/4in thick over the newspaper body. After final shaping it was allowed to dry for a week. Last, A+B epoxy putty (Pratley) was used to smooth the whale's skin. The lower mandible was constructed separately and contained forty-eight teeth made from bent wire covered with rawhide glue which shrank to give that old, puckered-up gum look. The upper jaw had sockets to accept only the lower teeth.

Before painting the whale in oils, the sky and sea were completed so that all three would have a common base colour, namely, Payne's Gray. His artistic sense told him that a sunny blue sky would do nothing to mirror the mood of a battle between man and whale. Instead, he chose a stormy, dark sky with a burst of golden light painted directly behind the whaleboat. This glow was eventually reflected on the whale's back and in the sea.

Turning to the sea, the first step was planning the direction of the waves and avoiding any ripples or waves which ran parallel to the edges of the case. It was built up of foam core, Celluclay, and Sobo Glue to smooth the water. This white craft adhesive is more flexible than Elmer's glue, and seems to stick where it is put rather than flow into puddles. A hole was left where the whale would break the surface, (also made at an angle to the case), and then it was put aside for a week to dry.

In the meantime he started work on the crew. The Airfix figures were World War II soldiers. These had to be stripped of their uniforms and heavy combat boots, then be re-dressed by contouring with A+B putty, Strathmore art board, paper towels and tissues. The crew's shoeless condition was reflected by carving toes on the feet.

At this point he broke off to paint the sky in various dark blues with the sunburst, then added dark green to those colours before painting the sea and the whale. The oils were allowed to dry for two weeks.

Returning now to the completion of the figures, each one received a coat of water-based acrylics and was painted exactly as he would appear when completed. This had several advantages. First, it allowed him to plan the figure's livery so that colours would enhance one another instead of clashing. Second, he learned the figure's anatomy and what the brush was going to do under the arms, in the crotch, around the folds of cloth, etc. Third, acrylics provided an excellent base for the final coat of artist's oil paints. Last, if the acrylics bled through the same shade of oil-based paint it would not be nearly so noticeable as when only grey or white primer was used. Highlights and shadows were added while applying the final coat of paint.

The component parts of the diorama were assembled by first gluing the whale into place with 5-Minute Epoxy. The hole in the sea was filled with more epoxy and Liquitex Gel Medium (a Pure acrylic polymer emulsion gel sold at all art supply stores). Liquitex Gel was employed throughout to form rough water and white water. These effects were easily achieved by agitating the gel with a toothpick as it set. After being allowed to dry overnight, the churning water was painted in pale blue and greens, with pure white and off-white whitecaps being added later.

The explosion of water as the whale thrust himself skyward was created by using thin strips of clear acetate or celluloid suspended from clamps and treated with 5-Minute Epoxy which was allowed to run down them to form teardrops at the bottom. They were mounted on the whale with more 5-Minute Epoxy and additional Liquitex Gel forming spume.

Four seagulls and an equal number of sharks were made for the diorama. Seagulls flying in a group told whalers where their prey would break surface, and sharks were attracted to the scene by the whale's blood. By painting a whaling ship in the background using the same colours as in the sky, not only was a viewer shown from where the courageous crew came, but extra depth was added to the diorama. At the horizon satin varnish was used to give the water a gentle glow, with three coats of high gloss varnish to produce the wet look as the eye moved forward.

HMS *Royal William*, 1719, in a choppy sea. Scale 1/75.
Victory photographs by Martin Konopacki Photography, Oak Part Ill.

After 1000 hours spread over six months he felt that he knew a little more about what whalers went through to get their prize. Looking at the dramatic scene he had created and entitled *Vengeance!*, he mused 'I can almost feel the pain'.

Sadly, on 20 May 1992, Larry Hubbell lost his battle with emphysema.

References

ANSELL, WILLITS D *The Whaleboat – A Study of Design, Construction & Use from 1850 to 1970.* Mystic: Mystic Seaport Museum, 1978.

CHURCH, ALBERT COOK *Whale Ships and Whaling.* New York, Bonanza Books, n/d.

Paintings by Thomas M Hoyne. ❏

Zachary and Elizabeth 1745-1759

A First Rate of the Imperial Russian Navy

by Karl Heinz Marquardt

Although warships taking part in eighteenth-century hostilities between Europe's western nations have proved popular subjects with present-day ship modellers, with on-going research producing excellent results, those involved in the equally bitter struggle of the two main rivals in the Baltic Sea has been markedly overlooked. Through the work of Frederik af H Chapman much is known about the various types of Swedish vessels of the second half of that century, but how much is known, especially by modelmakers, of their great adversary, the Russian fleet?

An enigma to many people since its foundation by Tsar Peter the Great in the eighteenth century, the Russian fleet fought many battles and produced good looking ships, some of which were built by English shipwrights who served, along with shipwrights from other nations, in Russia under several Tsars and Tsarinas. Russia had to defend frontiers both in the north and the south. In the north, the ships needed to oppose the Swedes were built in

Contemporary 1/32 scale model of the *Zachary & Elizabeth*, the third Russian 100-gun First Rate. The model was restored by the author in 1953/54.

busy Baltic Sea shipyards in places such as Kronstadt, Revel (modern-day Tallinn) and St Petersburg. The last named also produced about 260 skampavajs (half galleys), as well as all kinds of men-of-war during the Nordic wars. In the south, ships were built on the rivers Don and Dnieper at places like Voronezh, to counter Turkish supremacy in the Black Sea.

The flagship of the Baltic fleet, Russia's principal fleet since 1727, was a First Rate. During the 150-year span of the Russian sailing navy, only about twenty-eight of this type were built, and for most of that period only one of these wooden ships was in service at any one time. The first to be built was the 100-gun *Peter I & II*, construction of which began in 1723 under the direct auspices of Peter the Great, and was completed after his death in 1725. The second, a 110-gun ship named *Imperatritsa Anna* was built in 1737 by the English Master Shipwright P Brown. The 100-gun *Zachary and Elizabeth*, the subject of this article, was the third, being built by Dmitry Shcherbatchev, and launched in 1747. She was replaced in 1758 by another 100-gun ship, the *Sv Dmitry Rostovsky*, built by the English shipwright John Sutherland, who came to Russia from Holland in 1737 as an Assistant Master Shipwright. He launched his first ship in 1743, and his list of achievements included three 80-gun ships and several 66s.

For a long time the model described here was thought, incorrectly, to be of *Peter I & II*, and has been part of a collection at Castle Eutin/Holstein belonging to the Grand-Dukes and Dukes of Oldenburg since the early part of the nineteenth century. In attempting to identify this large model of a three-decker, one of two 200-year old models which I had the good fortune to restore in 1953-54, the four above named First Rates were considered carefully. At a scale of 1/32

Stern of the model after restoration.

(3/8in = 1ft) the model did not correspond to the specifications of the first two. Both ships were square tucked and, moreover, the *Imperatritsa Anna* carried 110 guns. The model and the *Zachary* not only have a round tuck in common, being the first 100-gun ship built with one, but the dimensions of the model matched that ship's known measurements.

Although deck arrangements and dimensions point to the first half of the eighteenth century, some of the external decorative features belong to an earlier period. The presence of round and rectangular gunport decorations, applicable only to late seventeenth- and early eighteenth-century English and Dutch ships, on the *Zachary* hint at construction and supervision by other than a Western shipwright.

The model is one of the few remaining from Russia's early fleet and is probably the only surviving fully-rigged model of a First Rate from the first half of the century. In the same collection is another three-decker with fourteen gun tiers from about the same period, an inferior halfmodel known as the frigate *Augsburg*.

The Master Shipwright of *Zachary and Elizabeth* must have been a remarkable man, being honoured in a similar fashion as af Chapman, who received the rank of Vice-Admiral in his later years. To be chosen to construct the fleet's flagship was indeed a distinction. We do not know much about the ships he constructed. His career began in 1715 when he joined the armed forces as a trainee. Peter I initiated a large shipbuilding programme on the Don and the Dnieper between 1722 and 1725 in preparation for another war with Turkey. Selected in 1723 as an apprentice shipwright for that programme, Shcherbatchev was sent to Voronezh on the River Don, earning his spurs as a junior shipbuilder in the construction of Peter's Black Sea fleet. Eleven years later he went as a shipbuilder to Revel on the Baltic Sea, and in 1735 was promoted to Major. A Colonel in 1740, he constructed the 66-gun ship *St Peter* between 1740-41 and the *Zachary and Elizabeth* between 1745-47. Having launched the latter, Dmitry Shcherbatchev became Chief-Surveyor Captain-Commander, and in 1757 received the rank of Major-General.

Construction of *Zachary and Elizabeth* commenced on 26 May 1745 at the Admiralty main yard at St Petersburg. She was launched about two years later on 12 October 1747 and served in the Kronstadt squadron of the Baltic Fleet. As she was not completed until after the second Swedish/Russian War, she only served in peacetime. When the Seven Year War began in 1756 she was found to be unfit for further service. She was built mainly of fir, and her hull having rotted away,

The author applying the final touches in 1954 to the restored model's rigging. A comparison between model and man gives a good indication of its size.

she was decommissioned in 1759, and soon after broken up at Kronstadt. The reasons for building ships from fir in this region were economic. Fir was more easily available than oak and the building costs were a mere third; however, a fir ship lasted only two-thirds of the time of an oak ship.

With the first two flagships named after the ruling monarchs of the time, the name *Zachary and Elizabeth* was a departure from that custom. It does not refer to the reigning Tsarina Elizabeth I but, by following another tradition, it honoured major saints in the orthodox church, the parents of John the Baptist.

Specification

Length over lower gundeck: 181ft
Breadth (not given as extreme): 51ft
Depth in hold: 21ft 9in
(These are Russian feet, which were similar to English feet. 1ft = 0.3048m)

ARMAMENT
Lower deck: 28 30pdrs
Middle deck: 28 18pdrs
Upper deck: 28 12pdrs
Quarterdeck: 12 6pdrs
Forecastle: 4 6pdrs

The model

When I commenced restoration work in 1953 the bad state of the model made it necessary to dismantle it almost completely.

Her worm-eaten planking, which in many places was not much more than paper-thin outer layers with wood-dust in between, had to be stabilised with an injection of worm-destroying and wood-dust binding substance, normally used for the preservation of worm-infected wood carvings. The rigging was carefully taken off and the decks were removed. Although the hull planking and framing had been stabilised, they were still in a very fragile state, and in between 50 per cent and 80 per cent of all the deck beams had to be replaced, being so worm-eaten that they could no longer give any strength to the hull. The same could be said for the decks where most of the timber had to be renewed. Adding these new timbers to the model considerably strengthened the hull and the more intricate and detailed work could commence. For example, the figurehead was cleaned and refitted, and some new pieces of side ornamentation had to be carved to replace that which had been broken off; twenty-eight missing guns were rebuilt, and about sixty pieces of missing stern decoration and balcony segments were added, as well as bad pieces of a former restoration replaced. Several of the internal bulkheads and other items were so worm-eaten that they had to be replaced.

After about four months of intensive work, every piece had been overhauled and was back in its place. A clean up of the original colours and paint work was still needed and once this had been completed, re-rigging could commence.

The rigging

The ravages of time had wreaked considerable damage to the rigging. Royals had been added in an earlier restoration, and the furled sails of

Starboard elevation with, below, detail of the inboard side of the upperdeck port bulwark. Other illustrations show the sections in greater detail. Drawing produced from restoration notes and sketches.

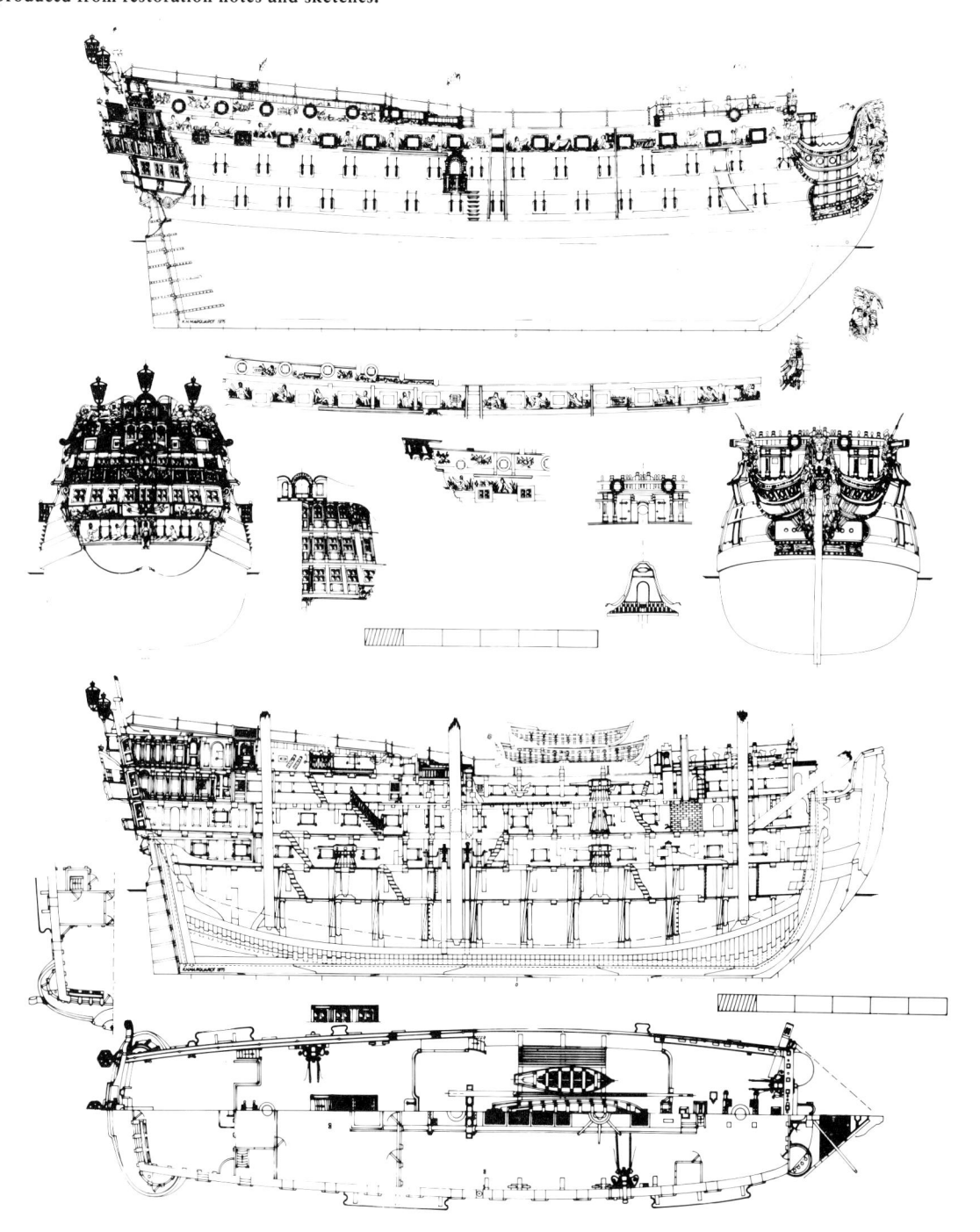

Sectional elevation based on restoration notes and sketches. Orlop deck, boats and boat beams, pumps added. The port side of the plan view shows coach, quarterdeck, main deck and forecastle. The starboard side is the plan of the main deck. To the left is a detail of the coach with cabin, lobby and stairway.

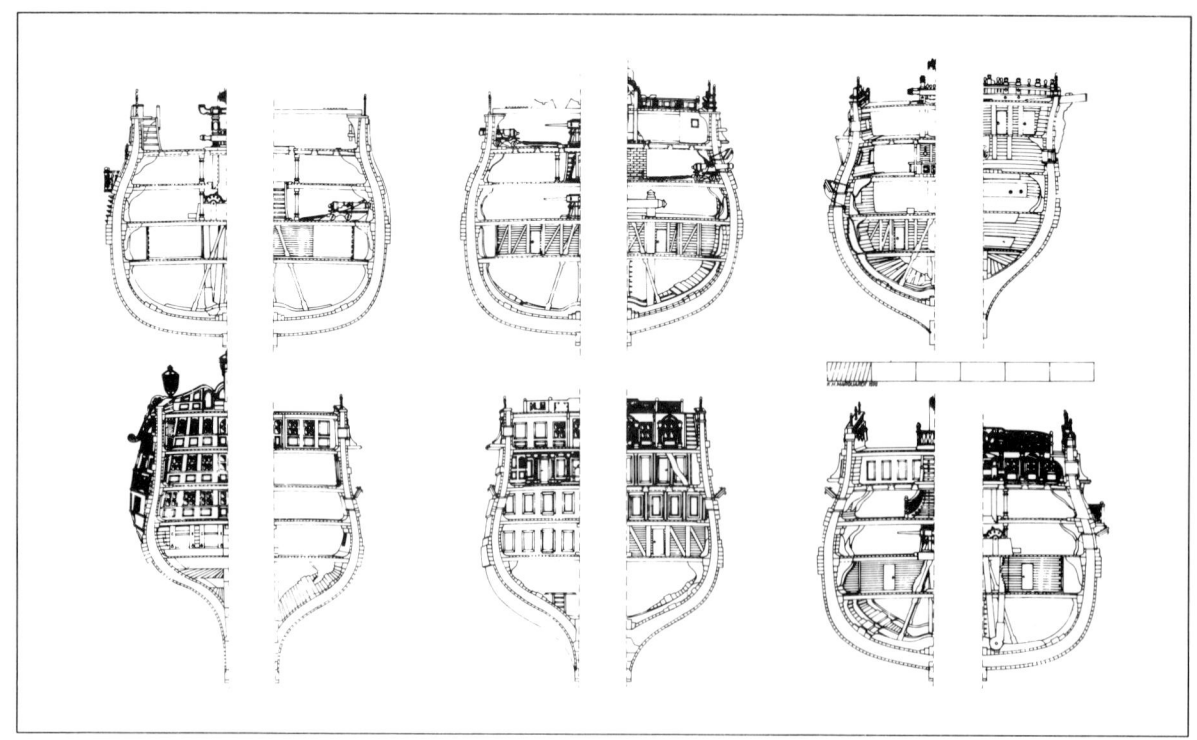

Various cross sections through the vessel.

these later yards proved to be part of a studdingsail and of the missing spritsail. Some iron mastbands had to be replaced, tops needed repair, and jibboom, main topmast and topgallant mast, beside other spars, had to be repaired or replaced. Three sails, spritsail-topsail, main topgallant sail and the mizzen were remade (hand stitched) from two-hundred-year-old linen.

The standing rigging was still good enough to be put back into place. However, much of the running rigging was very brittle, had been shortened during the earlier restoration (probably during the nineteenth century) and in many instances needed to be replaced. A quarter of all blocks and two anchors were missing. The restoration of the *Zachary and Elizabeth* at Castle Eutin/Holstein was completed after 7½ months in early 1954.

An essential part of dismantling such an intricate piece of art was to record the place of all the parts. Later on, an attempt was made to work up these notes, and to turn the rough pencil sketches, made during the work of restoration, into a set of ship model plans. It was, however, too big a task to complete at that time.

In 1975 the drawings were traced in ink on to drafting film, with the boats in place as well as the pumps, and the orlop deck added. Information about the type of pumps used on the ship was unobtainable, and so the British type of chain pump was included because of the English influence in Russian shipbuilding. A recent redrafting of some damaged plans led to a thorough counter-checking of the work, drawings, and measurements with several side view photographs of the model, and this revealed certain dimensional inaccuracies in the earlier notes. These errors in length, stem shape, sternpost rake, etc, meant that some details had been drawn too short, which reflected therefore on the overall length. They have been eliminated on a new sheer plan, which is thus slightly longer than the former and more correct. The new plans, in conjunction with the earlier drawings, should enable a good modelmaker to recreate a good likeness of that handsome ship of nearly 250 years ago.

Comparison of the exterior hull with British ships of that period reveals quite a few similarities. The width of the main wales and the black strake above, the 3ft-wide middle wale, the gun port placement and the roundhouses suggest English inspiration. Yet, while the ship's gun deck length is slightly larger than that of 100-gun ships of the 1745 Establishment, with the two other main dimensions correlating, other deck measurements would be more in line with the 1719 Establishment. However, these dimensions are only indirectly related to English Establishments, as they were instigated by Peter I in 1718. He gave shipbuilding specialists in his service (Englishmen amongst them) the task of laying down all major dimensions for the four Rates of 100-, 80-, 66-, and 54-gun ships

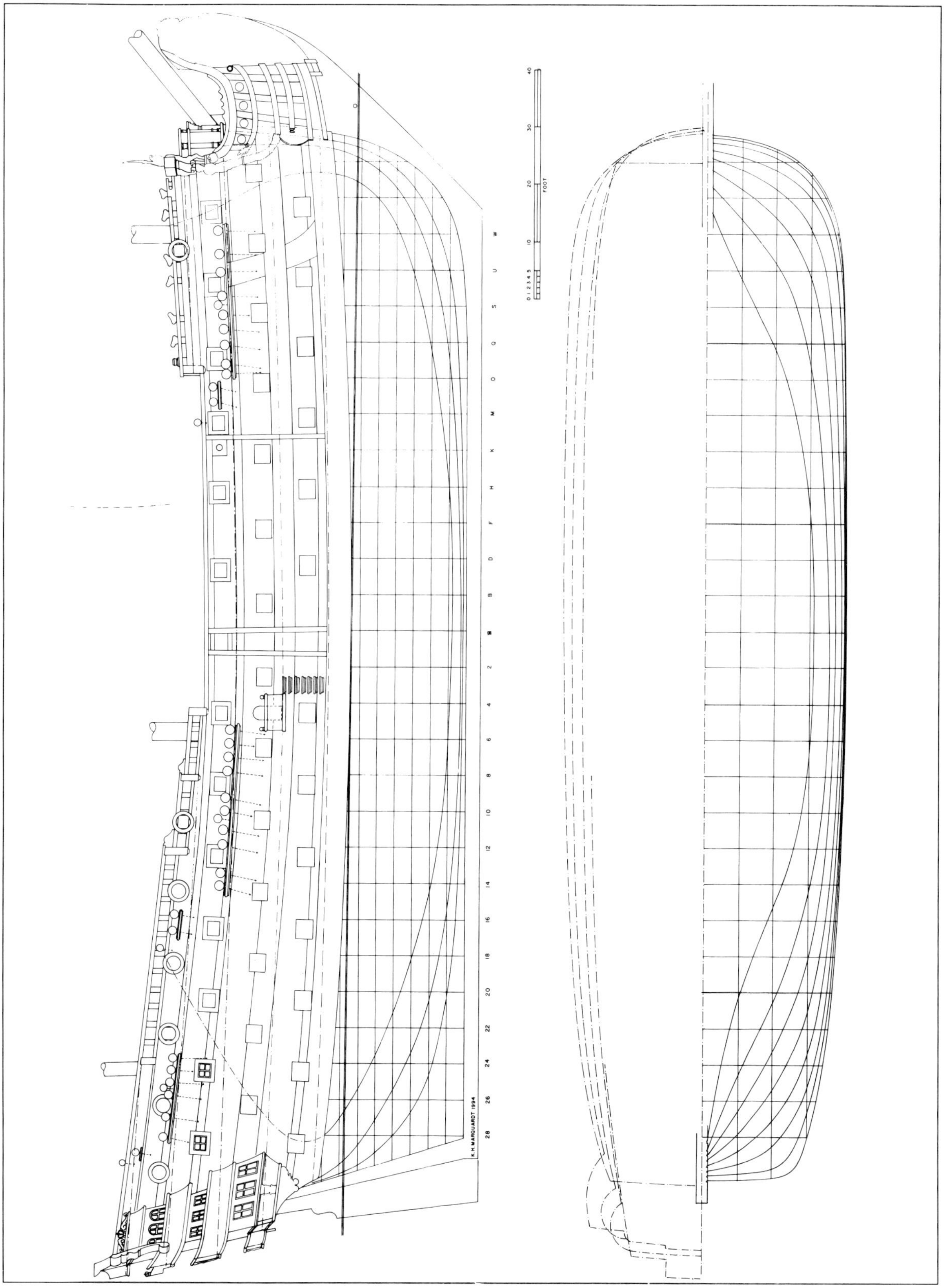
K.H.MARQUARDT 1994
FOOT

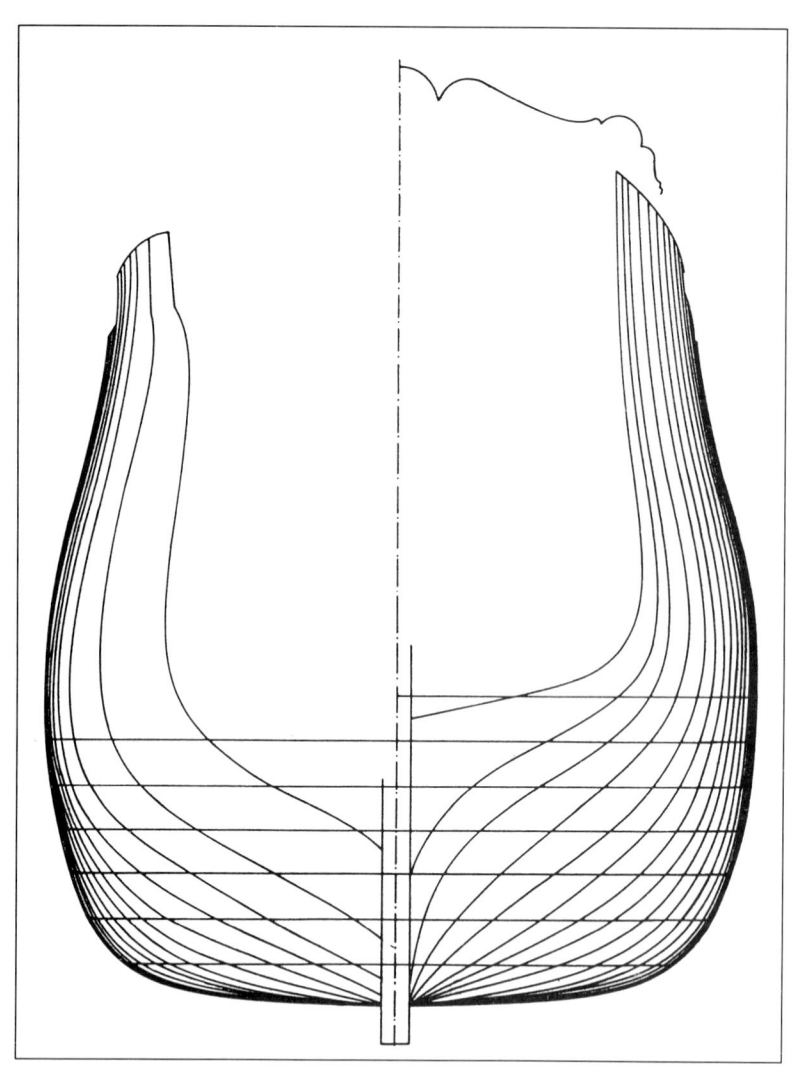

Mr Demidow from the Russian Central Naval Museum, be found as well on *Peter I & II*. They can be seen on a drawing of the Russian 80-gun ship *Saint Paul* of 1743, and more evidence may probably be gained from other Russian models and draughts from before the mid-century.

Painted decoration

The space between the round quarterdeck ports was painted red and filled with military emblems, such as shields with crossed flags, lintstocks, spears, guns etc, with the smaller sections showing intertwined ornamental scrolls. There was only a red coloured band above the sheer rail at forecastle deck level. Between the sheer and waist rails, mainly within the square decorations of the upper deck ports, was an azure blue band with bathing figures frolicking between reeds. None of these figures and emblems was duplicated from side to side: all were individual.

The entry port

Abreast of the mainmast on the middle deck, between the eighth and ninth gunports, was the entry port. The semicircular upper part of the entrance was flanked by two pillars, topped by an arch and bearing a crown, which in turn was balanced by two ornamental scrolls. The visible woodwork between pillars, arch and entry opening was carved masonry. The landing balustrade panels were of similar style carvings to the main stern gallery. Seven gangway steps reached down to the upper edge of the main wale. Skids were fitted fore and aft of the seventh, and aft of the fourth upper and lower ports. The chess-tree, normally on English ships of that period built into the foremost skid, was not evident and an embellished lead hole for the main tack, (a seventeenth-century relic) was placed below the sheer rail directly aft of the skid.

and for 32-gun frigates. Completed in 1724, these strictly enforced regulatory tables were written into law on 9 July 1725, soon after Peter's death, during the short reign of Catherine I, and were still in force until 1777.

On the model the channel arrangements, gun port decorations, figurehead shape and open stern galleries were of pre-1745 English design, but also show a strong Russian influence. There were fourteen gun ports in every full tier, with an additional four stern ports on the lowest tier. Rectangular carvings framed the upper deck ports. Such carvings were indicated by Anthony Deane in 1670 for English ships, and can still be seen on the picture of the Dutch 52-gun ship *Gertruda* of 1720, and the model of the Dutch East Indiaman *Den Ary* of 1725. Sliding windows were fitted into the last two of these ports. Carved wreaths, as on the quarterdeck and forecastle ports, left the English shipbuilding scene about 1710. They were, however, still used on Russian ships, as on the model of *Isaak – Victoria*, a 66-gun ship from 1719, built by J Ney (Russian name: Osip Naj), who was hired in 1698 by Peter I and was one of the first English master shipwrights in Peter's service. Round port decorations, apart from those on *Zachary and Elizabeth* could, according to

The original stern drawing reworked according to photographs and restoration notes.

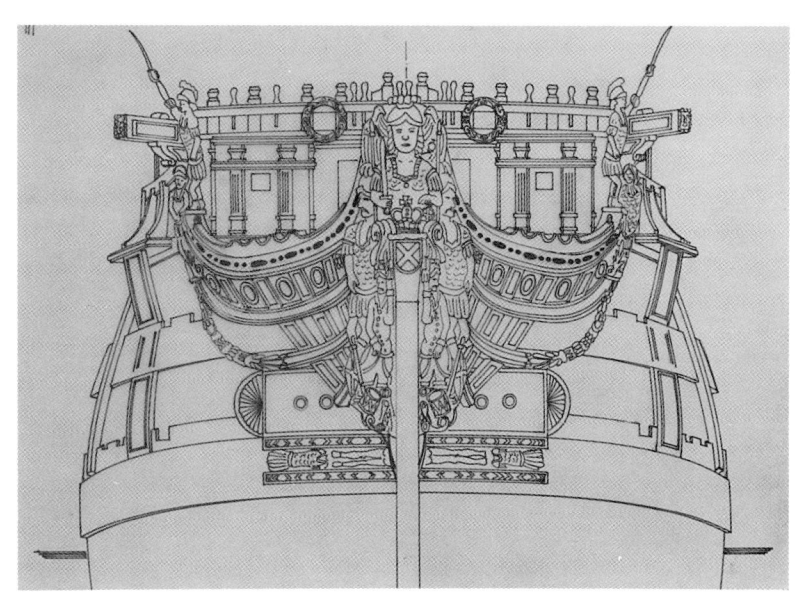

Enlargement of the figurehead and bow.

Forecastle and head detail.

Channels

The fore and main channels were situated below the upper deck, and the mizzen channel below the quarterdeck ports, as on similar large English ships pre-1745. Backstay securing points were: fore channel, forward of the fourth upper deck port; main channel, between the second and third quarterdeck port; mizzen channel, forward and above the last quarterdeck port. Fore and main channels were supported by three, and the mizzen channel by two, inverted timber knees. They stood much further apart than the average of 2½ft for English ships. An anchor lining was fitted to the fore channels. An item rarely found on models and on pictures are the stanchions for the arming cloths (fighting, or waist cloths). They were placed from stern to bow over the whole length of the ship, and these cloths of kersey would have protected the crew during close battle against snipers. This was a practice more commonly found during the seventeenth century, and on English ships, usually covering only quarterdeck and forecastle. The protruding timberheads at the forecastle were of the 'finger & thumb' kind, a type also used during the first half century on English ships.

Trying to narrow the time scale by identification based on style elements is virtually impossible. It underlined the point that this ship was not built by an Englishman, but by somebody who applied practically all the finishing details he had learned during his training by one of the shipwrights Peter the Great imported at the end of the seventeenth century, or by one of their students. He combined these with the contemporary practices developed during his life as a shipbuilder, as for example, with the vertical carvings of hances and drifts, which were again a peculiarity of late seventeenth-century ships.

The figurehead

The ship's head was dominated by an elaborate figurehead of a female ruler, probably symbolising Elizabeth I, with a sword in her right hand, while the left rests on an orb.

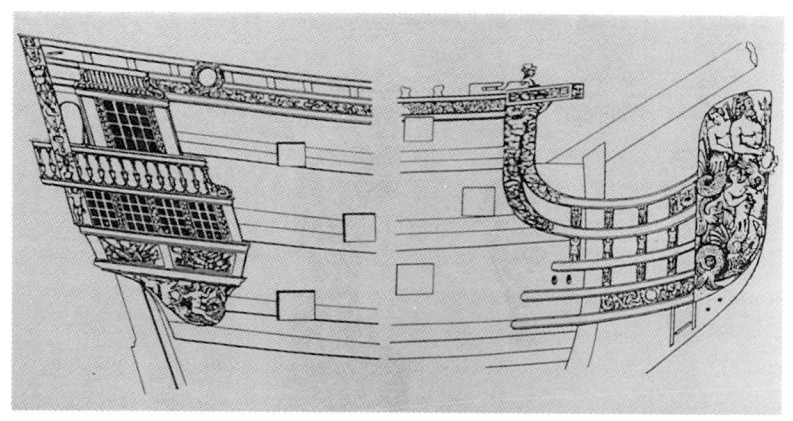

Dressed in ancient armour, she is standing in a stylised chariot or rostrum, embellished with a coat of arms and a crown. The double-headed Russian Eagle is spreading its wings behind her and two warriors in similar armour are on guard at a lower level. Beneath the rostrum are two kneeling men in eighteenth-century uniform, carrying the same plumed helmet as warriors and monarch and having an arm raised to pledge allegiance. The lower end of the figurehead is made up of flags, guns, drums and horns, with the aft side of the figure forming a sheltered room. Between the continuing cheeks of the head are carved naked (slain?) human figures and discarded ancient armour. The hawse holes above these are covered in English fashion with a naval hood, which ended in a semi-circle of sun rays.

The headrails

The headrails with their unadorned brackets in association with carved cathead supports and mainrail ending in a sword holding warrior, but with a full length combed false rail, are again a mixture of various styles, dating from the 1670s to the 1740s. Two roundhouses, with forward facing square windows, flanked the beakhead bulkhead, their roofs being joined athwartship by a 1½ft wide plank. A timber structure enclosed the knightheads and the bowsprit deck opening, enabling men to step from the forecastle on to the

Head and stern of the Russian 80-gun ship *St Paul* of 1742 (from an unknown source).

bowsprit. This enclosure had a larboard door and at both sides of it were doors in the beakhead bulkhead. Two wreathed chase ports were situated above.

The galleries

The general open gallery layout of the stern is comparable to similar Royal Navy ships until about *Royal George* in 1756. The stern's lower counter, pierced with four gun ports, was painted black-blue and decorated with bathing figures, with a double tailed nymph placed aft of the helm. A large balcony protruded from most of the upper counter, its centre decorated with a carved and gilded double-headed eagle and painted Russian flags, while its sides featured bathing figures on an azure blue background. Gilded carvings of seahorse riding Neptunes fill the visible parts of the upper counter proper. The lower tier of windows consists of eight windows and one double door, whilst the next two tiers have only six windows plus double doors. In contrast to the lower balcony the others are carried about to the galleries and appear elliptical in plan view. The middle balcony consisted of carved panels with another, larger double-headed eagle in its widened centre, with a shell-like, crowned and gilded canopy sheltering that section. Flat

and open balustrade panels alternate in the quarterdeck balcony enclosure, and over-lifesize carved sentinels guard the windows. All the balconies are supported by pillars between each of them. A short fourth balcony is part of the taffarel and approached from the poop by a door to port of the ensign staff. A bench on that balcony is placed under a central shell carving of the taffarel, and further carvings at each side are of nymphs attending sea deities. Three lanterns are carried at the stern and one at the main top. At the forward end of the poop is a carved footrail and stair at both sides leading up to that deck.

Internal decoration

The elaborate detailing does not stop at the exterior. Intricate deal panelling, interrupted by imitation pillars can be seen inside the coach and stateroom, while wardroom and gunroom have less complex panelling. Inside the stateroom a dais with a carved chair (throne) is placed in the centre of the forward bulkhead.

From the presence of the aforementioned canopy over the stateroom balcony entrance, one can conclude that the ship was designed with ceremonial functions by the monarch in mind.

The coach's interior consists of a large cabin, a lobby, rooms left and right of the lobby and stair alcoves on both sides. A door makes the larboard stairs accessible from the quarterdeck, while those on the starboard side are only for internal movement between coach and stateroom. Another interesting feature is built into the quarterdeck's main ladderway. The ladderway, with a decorated balustrade around, and the two adjoining companion gratings, with stanchions and ropes, had glazed sliding windows fitted between their deck beams to prevent rain from falling into the steerage.

The quarterdeck breastwork aft

of the mainmast was set back far enough to clear the main jeerbitts. Access from the upper deck was by way of decorated side gangways. Open gangways, 3ft wide, supported by five very strong timber knees and secured by the roped waist cloths and stanchions, joined the quarterdeck with the forecastle.

Recognisable accommodation features on the upper deck consist of, the stateroom, a lobby with two side rooms and the steerage, and two cabins, supposedly for boatswain and carpenter, under the after end of the forecastle. The after footrail of the forecastle was shaped similar to the quarterdeck breastwork, with the belfry in the centre. Both sides of the belfry were fitted with brackets for the spar timbers. Other peculiarities were a small stairway between hearth flue and fore jeer bitts, anchor davit shackles, and a small hatchway, covered with a diagonal grating, close to the beakhead bulkhead.

The stairway

An outstanding feature of the middle deck is a very wide, bell-shaped and balustraded stairway to the upper deck aft of the mainmast. All the fore and aft flights of stairs are slightly bell-shaped, whereas those leading athwartships are parallel. Furthermore, none of the side brackets, except for the major stairway just mentioned, are straight, but double curved. A brick-built hearth with enclosed galley is just aft of the foremast, an enclosure which also covers the mast itself and the bowsprit steps.

Port cathead, roundhouse, and structure over knightheads, taken during reconstruction.

Port side entry port. Note the painted strake with nude figures and leaf scrolls, the carved hances and drifts.

The capstans

On the lower deck were the main capstan, set between main and mizzen mast, and the warping capstan just below the double jeer capstan in the waist and on middle deck. They were all fitted with eight short bars, probably not of working length. Cable bitts were fitted to the lower deck, with a manger sited behind the hawse holes. Details drawn below the lower deck are not part of the actual model layout. They were only added to show how that space was used on the ship. Similarly, the athwartship beams for boat stowage have been added to show the way, the ship's boats were carried. Evidence for this lies in the reverse timber knees below the light fore and aft gangways, which are far too strong for the gangways alone.

Guns

The guns were all painted bronze and one, loose inside the model, carried an inked inscription '30 – 100 – pouts'. As 1 Russian pud = 16.38kg, the indicated weight would have been 1638kg, which is about the weight of a 12pdr gun. The inscription was probably pointing to the number of 12pdrs on board. Carriages were all built with a full bed, like those on Danish ships in the first half of the eighteenth century.

Further inscriptions were found in the rigging. One was a slip of paper, bearing the words *Voor Boom Lishel*, rolled into a fore studdingsail. The first two words are Dutch, while the third is close to the Dutch *Ly-zeilen*. Another was a name on a block *Do de Sjaw* or *Shaw*. With all these words written not in cyrillic but Roman letters one tends to think that the model could have been the work of western hands or of somebody educated in western languages.

Masting and rigging

In the masting and rigging are many items one would normally associate with ships of various nations. Differing from other ships of that period is the complete replacement of mast wooldings with iron bands, which took place on English ships about half a century later. Mast tops are of English type from between 1720 and 1745, while studding sail booms are fitted in the Dutch way, aft of the yards. The use

Forecastle, with belfry, guns, etc.
Photographs by the author

The coach during the course of reconstruction.

of stretchers, or square battens, above the shroud deadeyes is a feature usually associated with later ships.

The main and preventer stays were fitted with hearts, but in a Continental style, with both collars being led through a comb-cleat forward of the stem and joined forward of the foremast. The fore topmast stays were both set up with deadeyes to the bees (violin type). The bowsprit's cap was cocked like a French one, but made of iron, with the jibboom run on the starboard side beside the bowsprit.

A standard with a small cap stood at the outer end of the bowsprit, carrying the jackstaff. In comparison to English rigging, this was not an interim fitting of a jibboom together with a spritsail-topsail mast, but an advanced fitting in line with French and other Continental rigging.

Conclusions

For one who has an interest in eighteenth-century shipping and is visiting North Germany, a trip to the city of Eutin, with its castle near the Eutin lake, is a worthwhile detour from the beaten track. This large, not particularly well-known ship model, together with some others now shown, bears witness to a European land-locked nation's efforts more than two centuries ago to gain maritime access to its western neighbours. As one of the few 100-gun three-decked ships afloat anywhere in the world at the time, it also shows what a level of sophistication the first generation of trained indigenous shipwrights had achieved in an industry only four decades old. Also, it demonstrates well that the Russian Navy is not just a twentieth-century phenomenon, but is only a few years away from its 300th birthday. ❑